RECENTLY DISCOVERED SYSTEMS OF ENZYME REGULATION BY REVERSIBLE PHOSPHORYLATION

MOLECULAR ASPECTS OF CELLULAR REGULATION

VOLUME 1

General Editor

P. COHEN

Dundee

ELSEVIER/NORTH-HOLLAND BIOMEDICAL PRESS
AMSTERDAM · NEW YORK · OXFORD

RECENTLY DISCOVERED SYSTEMS OF ENZYME REGULATION BY REVERSIBLE PHOSPHORYLATION

Edited by

P. COHEN

1980

ELSEVIER/NORTH-HOLLAND BIOMEDICAL PRESS
AMSTERDAM · NEW YORK · OXFORD

ISBN Series: 0 444 80225 8
ISBN Volume: 0 444 80226 6

Publishers:
ELSVIER/NORTH-HOLLAND BIOMEDICAL PRESS
335 JAN VAN GALENSTRAAT, P.O. BOX 221
AMSTERDAM, THE NETHERLANDS

Sole distributors for the USA and Canada:
ELSEVIER/NORTH-HOLLAND INC.
52 VANDERBILT AVENUE
NEW YORK, N.Y. 10017

PRINTED IN THE NETHERLANDS

Editor's foreword

As the structures of metabolites and macromolecules, and the pathways by which they are synthesized and degraded have been elucidated, the main emphasis of biochemical research has progressively shifted towards problems of cellular organisation and control, a trend which has been particularly marked over the past five years.

While a number of series dealing with cellular control mechanisms started to appear in 1970s, most of these books were merely collections of reviews dealing with unrelated topics. This was perhaps inevitable since knowledge of control mechanisms at the molecular level had not advanced sufficiently for common underlying themes to have emerged. However the past few years have seen considerable advances and cellular regulation has now reached an exciting stage where unifying concepts are now linking areas of research which were previously thought of as being quite separate.

The purpose of *Molecular Aspects of Cellular Regulation* is not to produce a regular annual volume, but to publish occasional books on multidisciplinary topics which illustrate general principles of cellular regulation. The first volume concerns the role of protein phosphorylation in co-ordinating the control of intermediary metabolism. Volume 2 will deal with the molecular actions of bacterial toxins, viruses and interferon.

List of contributors

G.S. Boyd   Department of Biochemistry, University of Edinburgh Medical School, Teviot Place, Edinburgh EH8 9AG, Scotland

P. Cohen   Biochemistry Department, Medical Sciences Institute, University of Dundee, Dundee, DD1 4HN, Scotland

P.J. England   Department of Biochemistry, Medical School, University of Bristol, Bristol, BS8 1TD, England

L. Engström   Institute of Medical and Physiological Chemistry, Biomedical Center, University of Uppsala, Uppsala, Sweden

D.M. Gibson   Department of Biochemistry, Indiana University School of Medicine, Indianapolis, IN 46223, U.S.A.

A.M.S. Gorban   Department of Biochemistry, University of Edinburgh Medical School, Teviot Place, Edinburgh EH8 9AG, Scotland

D.G. Hardie   Biochemistry Department, Medical Sciences Institute, University of Dundee, Dundee, DD1 4HN, Scotland

T. Hunt   Department of Biochemistry, University of Cambridge, Cambridge, England

T.S. INGEBRITSEN   Department of Biochemistry, Indiana University School of Medicine, Indianapolis, IN 46223, U.S.A. Present address: Biochemistry Department, Medical Sciences Institute, University of Dundee, Dundee DD 4HN, Scotland

D.P. LEADER   Department of Biochemistry, University of Glasgow, Glasgow G12 8QQ, Scotland

H.R. MATTHEWS   Department of Biological Chemistry, University of California at Davis, CA 95616, U.S.A.

H.G. NIMMO   Department of Biochemistry, University of Glasgow, Glasgow G12 8QQ, Scotland

Contents

Chapter 6.   The hormonal control of triacylglycerol synthesis by H.G. Nimmo   *135*

Chapter 7.   Protein phosphorylation in the regulation of muscle contraction, by P.J. England   *153*

CHAPTER 1

Well established systems of enzyme regulation by reversible phosphorylation

PHILIP COHEN

1. Discovery of enzyme regulation by phosphorylation–dephosphorylation

Although it has been known for almost a hundred years that proteins contain covalently bound phosphorous, it is only since the discovery of enzyme regulation by reversible phosphorylation that interest in protein phosphorylation has gathered momentum (Figure 1). In 1956, Krebs and Fischer [1] discovered that glycogen phosphorylase the rate limiting enzyme in glycogenolysis could be converted from a dephosphorylated 'b'-form whose activity was dependent on the allosteric activator 5'-AMP to a phosphorylated 'a'-form which was largely active in the absence of 5'-AMP. In 1959 the same workers demonstrated that phosphorylase kinase, the enzyme which converted phosphorylase b to a was itself an 'interconvertible' enzyme which could exist in a low activity dephosphorylated state or a high activity phosphorylated state [2]. The third enzyme shown to be regulated by this mechanism was also in the field of glycogen metabolism. In 1964 Joseph Larner and his associates showed that glycogen synthase the rate limiting enzyme in glycogen synthesis could be converted from a dephosphorylated form of high activity to a phosphorylated form which required that allosteric activator glucose-6P for activity [3].

The idea that this mechanism of regulation might exist in other systems was slow to take root. It was only after the discovery of cyclic AMP-dependent protein kinase in 1968 by Walsh, et al. [4], also as a result of studies of the control of glycogen metabolism by hormones, that the field started to move rapidly and there are now some 25 enzymes whose activities have been demonstrated to be regulated by phosphorylation–dephosphorylation in vitro (Figure 1). However, the number of phosphoproteins (as opposed to phospho-

Cohen (ed.) Recently discovered systems of enzyme regulation by reversible phosphorylation
© Elsevier/North-Holland Biomedical Press, 1980

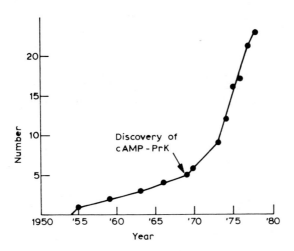

Figure 1. Enzymes reported as undergoing regulation by phosphorylation–dephosphorylation. (Reproduced by permission of Edwin G. Krebs.)

enzymes) that have been identified is now greater than one hundred.

The past five years have established protein phosphorylation as the major general mechanism by which intracellular events in mammalian tissues are controlled by nervous and hormonal stimuli. The purpose of this book is to describe some of these newly discovered systems of enzyme regulation by phosphorylation–dephosphorylation which have been instrumental in developing our current understanding of how the major pathways of intermediary metabolism are controlled in a co-ordinated manner in response to physiological stimuli. It will become apparent from the book that the regulation of carbohydrate metabolism, lipid, steroid and protein synthesis, and chromosome condensation show striking similarities and the general themes that will emerge are discussed in Chapter 11.

Although this book is concerned primarily with newly discovered systems, it is apparent from the foregoing discussion that the first three enzymes shown to be regulated by phosphorylation–dephosphorylation were those concerned with the regulation of glycogen metabolism in mammalian skeletal muscle, and this system continues to acts as the model to which all others are compared. It is therefore important the summarize briefly our current understanding of this system and its implications for other cellular processes which respond to neural and hormonal stimuli.

2. Regulation of glycogen metabolism in mammalian skeletal muscle

Glycogen metabolism in muscle is regulated by the hormones adrenaline and insulin as well as by the contractile state of the tissue. Electrical excitation of muscle or the release of adrenaline into the circulation stimulate glycogenolysis, while insulin promotes glycogen synthesis. These stimuli act by changing the activities of glycogen phosphorylase and glycogen synthase the rate limiting enzymes in glycogenolysis and glycogen synthesis respectively. The interrelationships between the protein kinases and phosphorylated proteins involved in this system are summarized in Figure 2. Progress up to the end of 1977 has been described in two reviews [5,6].

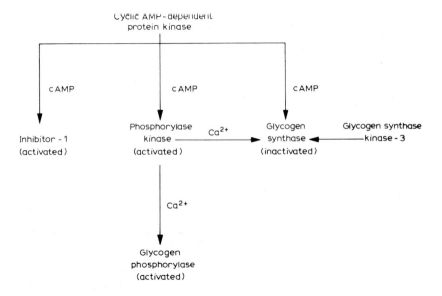

Figure 2. Interrelationship between the protein kinases and phosphorylated proteins involved in the regulation of glycogen metabolism in mammalian skeletal muscle.

The interaction of adrenaline with its β-receptors on the outer surface of the plasma membrane leads to the activation of adenylate cyclase located on the inner surface on the membrane. This elevates the intracellular level of cyclic AMP leading to the activation of cyclic AMP-dependent protein kinase. Cyclic AMP-dependent protein kinase possesses the structure R_2C_2, where R, the regulatory subunit, binds cyclic AMP and C is the catalytic subunit. The

4

binding of cyclic AMP to the inactive R_2C_2 complex causes it to dissociate thereby liberating the active catalytic subunit.

$$\underset{\text{(inactive)}}{R_2C_2} + cAMP \rightleftharpoons R_2 - cAMP + \underset{\text{(active)}}{2C}$$

The catalytic subunit controls glycogen metabolism by phosphorylating the three proteins inhibitor-1 (see below), phosphorylase kinase and glycogen synthase. The phosphorylation of phosphorylase kinase and glycogen synthase increases and decreases the activities of these two enzymes respectively. Thus the two opposing pathways of glycogenolysis and glycogen synthesis can be regulated in a synchronous manner in response to adrenaline.

Electrical excitation of muscle causes calcium ions to be released from the sarcoplasmic reticulum into the cytoplasm. The calcium ions not only activate actomyosin ATPase and initiate muscle contraction, but also activate phosphorylase kinase which is almost completely dependent on this divalent cation. The activation of phosphorylase kinase promotes the conversion of phosphorylase b to a, thereby stimulating glycogenolysis to provide the energy which sustains muscle contraction.

Recently, phosphorylase kinase has been shown to phosphorylate glycogen synthase and to decrease its activity [7–11]. Thus the two opposing pathways of glycogenolysis and glycogen synthesis, as well as the muscle contractile apparatus itself, can be controlled in a co-ordinated manner in response to neural stimulation of the tissue.

Phosphorylase kinase possesses the structure $(\alpha\beta\gamma\delta)_4$. The α- and β-subunits are the components phosphorylated by cyclic AMP-dependent protein kinase and the γ-subunit appears to be the catalytic subunit [12]. Recently, the δ-subunit was identified as the calcium binding protein termed calmodulin [13–15]. This protein is also a subunit of a number of other calcium dependent enzymes and will be discussed further in Chapter 7 and in Chapter 11. It is becoming apparent that calmodulin is the major cytoplasmic calcium receptor in eukaryotic cells and that it plays a role analogous to that of the regulatory subunit of cyclic AMP-dependent protein kinase.

The amino acid sequence of calmodulin has demonstrated a 50% identity with troponin-C [16], the protein which confers calcium sensitivity to the actomyosin ATPase reaction in the muscle contractile apparatus. Troponin-C and calmodulin both bind four calcium ions per mole with affinities in the micromolar range [17]. These findings allow one to start to visualize at a molecular level how the processes of muscle contraction and glycogen metabolism are so closely synchronized.

Glycogen synthase is also phosphorylated by at least one further enzyme which has been termed glycogen synthase kinase-3 [18,19], but its mechanism of regulation is unknown. Although phosphorylation by each of the three glycogen synthase kinases decreases the activity of glycogen synthase, and makes the activity more dependent on glucose-6P, different phosphorylation sites on the enzyme are involved (Table 1). Cyclic AMP-dependent protein kinase preferentially phosphorylates two serine residues termed sites 1a and 1b. Site 1a is phosphorylated and dephosphorylated much more rapidly than site 1b and appears to be responsible for the activity changes that accompany phosphorylation. Phosphorylase kinase phosphorylates a single serine located seven amino acids from the N-terminus of glycogen synthase (site 2) [9], while glycogen synthase kinase-3 phosphorylates three serine residues (sites 3a, 3b, 3c) all located in the same nine amino acid segment of the polypeptide chain [19]. Phosphorylation of all six serine residues by the three protein kinases is necessary to achieve a completely inactive enzyme [18].

Phosphorylase kinase, like glycogen synthase, is phosphorylated by cyclic AMP-dependent protein kinase at two serine residues, one on the α-subunit and one on the β-subunit (Table 1). The β-subunit is phosphorylated most rapidly and correlates with the rise in activity. The phosphorylation of the α-subunit has not so far been shown to have a direct effect on the activity, but its phosphorylation appears to stimulate the rate at which the β-subunit can be dephosphorylated.

Glycogen synthase and phosphorylase kinase were the first two examples of enzyme regulation by multisite phosphorylation, and the phosphorylation of an enzyme at more than one site by more than one kinase is turning out to be the rule rather than the exception. The mitochondrial pyruvate dehydrogenase complex, another well established example of enzyme regulation by reversible phosphorylation is also phosphorylated at several sites during its inactivation by pyruvate dehydrogenase kinase [20,21]. Further examples of this phenomenon will be found throughout the book. The covalent attachment of a second (or further) phosphate(s) can increase the potential of an enzyme for regulation considerably. Additional phosphorylations can alter the kinetic parameters in ways that either differ from or extend the effects produced by the primary phosphorylation. Furthermore additional phosphate groups can be inserted or removed by different protein kinases and protein phosphatases which can then act as separate targets for metabolic control.

Phosphorylase a, phosphorylase kinase (β-subunit) and glycogen synthase can be dephosphorylated and inactivated by a single enzyme which has been termed protein phosphatase-1 [6]. This enzyme can therefore carry out each of the dephosphorylations which inhibit glycogenolysis or which activate glycogen synthesis (Figure 3). The α-subunit of phosphorylase kinase is

TABLE 1

Amino acid sequences at the phosphorylation sites of the enzymes of glycogen metabolism (see [6,10,19])

(a) Sites phosphorylated by cyclic AMP-dependent protein kinase
(The roles of the basic amino acids just N-terminal to the phosphorylated residues in determining the specificity of this enzyme are discussed in Chapter 11).

Protein	Sequence
Phosphorylase kinase (α-subunit)	phe-arg-arg-leu-ser(P)-ile-thr-glu-ser
Glycogen synthase (site 1a)	arg-arg-ala-ser(P) lys
Phosphorylase kinase (β-subunit)	ala-arg-thr-lys-arg-ser-gly-ser(P)-val-lys-pro-leu-lys
Glycogen synthase (site 1b)	gly-gly-ser-lys-arg-ser-asn-ser(P)-val-asp-thr-ser-ser
Inhibitor-1	ile-arg-arg-arg-arg-pro-thr(P)-pro-ala-thr

(b) Sites phosphorylated by phosphorylase kinase
(The numbers indicate distances from the N-terminus of each enzyme. Residues in identical positions are marked by a full time and conservative differences by a broken line).

Protein	Sequence
Glycogen phosphorylase	glu-lys-arg-lys-gln-ile-ser(P)-val-arg-gly-leu-ala-gly-val-glu [10] [14] [20]
Glycogen synthase (site 2)	pro-leu-ser-arg-thr-leu-ser(P)-val-ser-ser-leu-pro-gly-leu-glu [5] [7] [10] [15]

(c) Sites phosphorylated by glycogen synthase kinase-3

Protein	Sequence
Glycogen synthase (sites 3a, 3b, 3c)	arg-tyr-pro-arg-pro-ala-ser(P)-val-pro-pro-ser(P)-pro-ser-leu-ser(P)-arg [3a] [3b] [3c]

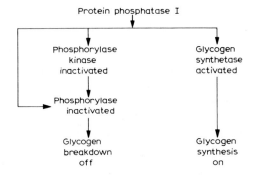

Figure 3. Role of protein phosphatase-1 in the control of glycogen metabolism.

dephosphorylated by a distinct enzyme which has been termed protein phosphatase-2 [6]. This enzyme also has some phosphorylase phosphatase and glycogen synthase phosphatase activity.

Protein phosphatase-1 is inhibited by a heat stable protein in skeletal muscle termed inhibitor-1. This protein is only an inhibitor of protein phosphatase-1 after it has been phosphorylated on a unique threonine residue by cyclic AMP-dependent protein kinase (Table 1, Figure 2). Inhibitor-1 is therefore the protein which allows cyclic AMP-dependent protein kinase to inactivate protein phosphatase-1 (Figure 4). It therefore provides a mechanism for amplifying the effects of cyclic AMP, and for exerting a tight control over several protein kinase/protein phosphatase cycles. Inhibitor-1 has been shown to be phosphorylated in vivo, and its degree of phosphorylation is markedly enhanced by adrenaline [22].

Interestingly, inhibitor-1 does not inhibit its own dephosphorylation even at concentrations 1000-fold higher than those required to prevent the dephosphorylation of the other substrates of protein phosphatase-1. There is therefore no need to invoke the existence of a separate enzyme for the dephosphorylation of inhibitor-1. However other inhibitor-1 phosphatases may well exist, and protein phosphatase-2 has been shown to dephosphorylate and inactivate inhibitor-1 in vitro [23].

Protein phosphatase-1 clearly has a rather broad specificity since it dephosphorylates a number of sites that are phosphorylated by several protein kinases (Figure 3, Table 1). It will however become apparent from this book that protein phosphatase-1, and also inhibitor-1, are likely to play an even wider role in the control of different cellular processes. This idea will be discussed further in the concluding chapter.

8

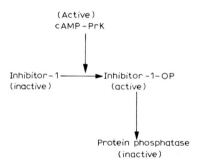

Figure 4. Cyclic AMP-dependent protein kinase (cAMP-PrK) inactivates protein phosphatase-1 through the phosphorylation of inhibitor-1.

3. Cyclic AMP and hormone action

Although cyclic AMP and cyclic AMP-dependent protein kinase were originally discovered during studies of the regulation of glycogen metabolism by adrenaline, it is now well established that these molecules play an important role in the regulation of many cellular processes. Cyclic AMP and cyclic AMP-dependent protein kinase are present at similar concentrations in all mammalian tissues, even those tissues where glycogen metabolism is of very minor importance [24]. Furthermore a variety of hormones with diverse physiological effects in a wide range of tissues can all elevate the intracellular concentration of cyclic AMP. These findings have given rise to the idea shown in Figure 5 [25]. According to this hypothesis the specificity of a hormone is determined by whether the receptor for that hormone is present on the outer membrane of the target cells and which physiological substrates for cyclic AMP-dependent protein kinase are present within those cells. If the receptor is present, then the hormone–receptor interaction leads to the activation of adenylate cyclase, elevation of cyclic AMP and activation of cyclic AMP-dependent protein kinase. The protein kinase then phosphorylates whatever substrate proteins happen to be present within those cells. As a result the biological activities of the protein substrates become modified, leading to an alteration in the physiology of the cells.

One obvious prediction of this model is that a very large number of substrates for cyclic AMP-dependent protein kinase must exist in different cells in order to explain the great diversity of action of hormones. Several new examples of enzymes which are regulated by cyclic AMP-dependent protein will be introduced throughout this book.

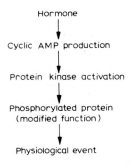

Figure 5. Hypothesis for the hormonal regulation of cellular functions working through cyclic AMP [25].

Edwin Krebs originally proposed a number of criteria that should be met before a protein could be said to serve as a *physiological* substrate for cyclic AMP-dependent protein kinase in vivo [26]. These criteria were recently redefined and made more stringent to take into account recent developments such as multisite phosphorylations [5]. The various enzyme systems described in this book should therefore be evaluated critically with the following criteria in mind:

(1) A protein substrate for cyclic AMP-dependent protein kinase should exist which bears a functional relationship to the process mediated by cyclic AMP. The rate of phosphorylation of that protein, in its native state, should be adequate to account for the speed at which the process occurs in vivo in response to cyclic AMP.

(2) The function of the protein should be shown to undergo a reversible alteration in vitro by phosphorylation and dephosphorylation, catalysed by cyclic AMP-dependent protein kinase and a protein phosphatase.

(3) A reversible change in the function of the protein should occur in vivo in response to cyclic AMP.

(4) Phosphorylation of the protein should occur in vivo in response to a hormone at the same site(s) phosphorylated by cyclic AMP-dependent protein kinase in vitro.

It is, however, important to emphasize that cyclic AMP-dependent protein kinase represents just one of a large number of different classes of protein kinase. A few protein kinases can be classified according to their mechanism of regulation, such as calcium-calmodulin dependent protein kinases (Chapters 7 and 11), cyclic GMP-dependent protein kinase [27, 28] and double stranded RNA-dependent protein kinase (Chapter 9). However, the

10

molecules which regulate most types of protein kinase have yet to be identified. The metabolic role of these co-called cyclic AMP-*independent* protein kinases is one of the most challenging problems still facing this field. This book will show that these enzymes accompany the action of cyclic AMP-dependent protein kinase in almost every control system. The possibility that they underlie the mechanism of action of hormones which do not work through cyclic AMP, such as insulin, will be considered in Chapter 11.

References

1. Krebs, E.G. and Fischer, E.H. (1956) Biochim. Biophys. Acta, 20, 150–157.
2. Krebs, E.G., Graves, D.J. and Fischer, E.H. (1959) J. Biol. Chem., 234, 2867.
3. Friedman, D.L. and Larner, J. (1963) Biochemistry, 2, 669–675.
4. Walsh, D.A., Perkins, J.P. and Krebs, E.G. (1968) J. Biol. Chem., 243, 3763–3765.
5. Nimmo, H.G. and Cohen, P. (1977) Adv. Cyc. Nuc. Res., 8, 145–266.
6. Cohen, P. (1978) Curr. Top. Cell. Reg., 14, 117–196.
7. Roach, P.J., DePaoli-Roach, A.A. and Larner, J. (1978) J. Cyc. Nuc. Res., 4, 245–257.
8. DePaoli-Roach, A.A., Roach, P.J. and Larner, J. (1979) J. Biol. Chem., 254, 4212–4219.
9. Soderling, T.R., Srivastava, A.K., Bass, M.A. and Khatra. B.S. (1979) Proc. Natl. Acad. Sci. USA, 76, 2536–2540.
10. Embi, N., Rylatt, D.B. and Cohen, P. (1979) Eur. J. Biochem., 100, 339–347.
11. Walsh, K.Y., Millikin, D.M., Schlender, K.K. and Reimann, E.M. (1979) J. Biol. Chem., 254, 6611–6616.
12. Skuster, J.R., Jessi Chan, K.F. and Graves, D.T. (1980) J. Biol. Chem., 255, 2203–2210.
13. Cohen, P., Burchell, A., Foulkes, J.G., Cohen, P.T.W., Vanaman, T.C. and Nairn, A.C. (1978) FEBS Lett., 92, 287–293.
14. Cohen, P., Picton, C. and Klee, C.B. (1979) FEBS Lett., 104, 25–30.
15. Shenolikar, S., Cohen, P.T.W., Cohen, P., Nairn, A.C. and Perry, S.V. (1979) Eur. J. Biochem., 100, 329–337.
16. Vanaman, T.C., Sharief, F. and Watterson, D.M. (1977) in: Calcium Binding Proteins and Calcium Funtion (Wasserman et al., eds.), pp. 107–116, Elsevier/North Holland, Amsterdam, New York.
17. Wolff, D.J., Poirier, P.G., Brostrom, C.O. and Brostrom, M.A. (1977) J. Biol. Chem., 252, 4108–4117.
18. Embi, N., Ryatt, D.B. and Cohen, P. (1980) Eur. J. Biochem., 107, 519–527.
19. Rylatt, D.B., Aitken, A., Bilham, T., Condon, G.D., Embi, N. and Cohen, P. (1980) Eur. J. Biochem., 107, 529–537.
20. Yeaman, S.J., Hutcheson, E.T., Roche, T.E., Pettit, F.H., Brown, J.R., Reed, L.J., Watson, D.C. and Dixon, G.H. (1978) Biochemistry, 17. 2364–2370.
21. Teague, W.M., Pettit, F.H., Yeaman, S.J. and Reed, L.J. (1979) Biochem. Biophys. Res. Commun., 87, 244–252.
22. Foulkes, J.G. and Cohen, P. (1979) Eur. J. Biochem., 97, 251–256.
23. Nimmo, G.A. and Cohen, P. (1978) Eur. J. Biochem., 87, 353–365.
24. Kuo, J.F. and Greengard, P. (1969) Proc. Natl. Acad. Sci., USA, 64, 1349–1355.
25. Krebs, E.G. (1972) Curr. Top. Cell. Reg., 5, 99–133.
26. Krebs, E.G. (1973) Endocrinology, Proceedings of the 4th International Congress, pp. 17–29, Excepta Medica, Amsterdam.
27. Lincoln, T.M., Dills, W.L. and Corbin, J.D. (1977) J. Biol. Chem., 252, 4269–4275.
28. Gill, G.N., Walton, G.M. and Sperry, P.J. (1977) J. Biol. Chem., 252, 6443–6449.

CHAPTER 2

Regulation of liver pyruvate kinase by phosphorylation–dephosphorylation

LORENTZ ENGSTRÖM

1. Introduction

Pyruvate kinase catalyzes the last, irreversible reaction of glycolysis, in which pyruvate and ATP are formed from phosphoenolpyruvate (= PEP) and ADP. During gluconeogenesis PEP is formed from pyruvate via oxalacetate in two other reactions catalyzed by pyruvate carboxylase and PEP carboxykinase, respectively. In mammals, these enzymes are mostly located in the principal gluconeogenetic organs, i.e., the liver and kidney [1]. In the liver the maximal activity of pyruvate kinase is very high (50 μmol of substrate transferred per min per g wet weight) as compared with that of both gluconeogenesis (1.7–1.8 μmol/min/g) and the pyruvate carboxylase and the PEP carboxykinase reactions, which are both about 6.7 μmol/min/g [1]. The activity of liver pyruvate kinase therefore has to be carefully regulated. In particular, the enzyme must be almost inactive during gluconeogenesis in order to avoid too great a degree of substrate cycling.

Hepatocytes contain only the L-type pyruvate kinase, while the enzyme of other liver cells is of the A (sometimes termed K_1 or M_2) type [2,3]. The concentration of the L-type pyruvate kinase protein in the liver is increased by insulin and a carbohydrate-rich diet and is decreased by fasting [4–6]. The activity of the enzyme is stimulated allosterically by fructose 1, 6-diphosphate (= Fru-1,6-P_2) and is inhibited by ATP and certain amino acids [5,7]. The intracellular concentration of *free* Fru-1,6-P_2 has been regarded as the most important factor for rapid regulation of the liver pyruvate kinase activity [5,8,9].

Glucagon and cyclic 3′,5′-AMP (= cAMP) activate gluconeogenesis in the liver, with an increase of the PEP to pyruvate concentration ratio [10, 11]. Since liver pyruvate kinase is not saturated with PEP in vivo, an increase in the concentration of this substrate would lead to enhanced enzyme activity if

Cohen (ed.) Recently discovered systems of enzyme regulation by reversible phosphorylation
© *Elsevier/North-Holland Biomedical Press, 1980*

it were not decreased by other means.

The physiological actions of glucagon are mediated by an increased intracellular concentration of cAMP [12]. Since the only known effect of cAMP in mammalian cells is to stimulate specific protein phosphorylation [13], it seemed reasonable to assume that under physiological conditions liver pyruvate kinase might be phosphorylated by cAMP-stimulated protein kinase, with a concomitant inhibition of the pyruvate kinase activity. Such an assumption was supported by the finding of Herman's group that the rat enzyme in vivo is partially inactivated within a few minutes after an intravenous injection of glucagon. Insulin, on the other hand, causes a rapid increase of the enzyme activity [14–17].

Highly purified liver pyruvate kinase from the rat and pig has been found to be a substrate of cAMP-stimulated protein kinase [18,19]. Up to 1 mol of phosphate can be bound to each subunit of the tetrameric enzyme. Isolation of a single peptic [^{32}P]phosphopeptide from the ^{32}P-labelled pig [20] or rat [21] enzyme has shown that a specific seryl residue of the enzyme is phosphorylated. The specificity of the reaction is further supported by the fact that neither the M-type enzyme from pig or rabbit muscle [22, 23], nor the A-type from pig kidney is phosphorylated by the protein kinase [22].

It, therefore seems probable that liver pyruvate kinase belongs to the group of enzymes whose activity is regulated by phosphorylation under the influence of cAMP. According to Krebs the establishment of such phosphorylation requires fulfilment of the following criteria [24]: (1) The cell type involved should be shown to contain a cAMP-stimulated protein kinase. (2) A phosphorylatable protein substrate bearing a functional relationship to the cAMP-mediated process should be identified. (3) It should be shown that phosphorylation of this substrate alters its function. (4) A phosphoprotein phosphatase which can reverse the phosphorylation process should be shown to exist. (5) It should be demonstrated that the protein substrate is modified in vivo in response to cAMP.

Nimmo and Cohen have recently modified these criteria [25]. Thus, they consider that the first and fourth criteria are not relevant, because of the ubiquitous distribution of cAMP-stimulated protein kinase and phosphoprotein phosphatase activities. With regard to the second criterion they point out that the rate of phosphorylation of the protein in its native state should be adequate to account for the speed at which the process occurs in vivo in response to cAMP. Nimmo and Cohen also state that a reversible change in the function of the protein should occur in vivo in response to cAMP. In addition, phosphorylation of the protein should occur in vivo in response to a hormone at the same *site* which is phosphorylated by cAMP-stimulated protein kinase in vitro.

In view of the cAMP-stimulated increase of PEP formation from pyruvate during gluconeogenesis [10,11], it is evident that liver pyruvate kinase bears a functional relationship to a cAMP-mediated process.

In this article results will be reviewed showing that mammalian liver pyruvate kinase in all probability belongs to the group of enzymes whose activity is regulated by a reversible phosphorylation–dephosphorylation according to the criteria of Krebs and of Nimmo and Cohen [24,25]. The structure of the phosphate-accepting site of pyruvate kinase will be described, as well as experiments demonstrating that synthetic peptides with amino acid sequences from this site are useful tools for studies of the protein kinase and phosphoprotein phosphatase reactions. The role of the phosphorylation of pyruvate kinase in the regulation of glycolysis and gluconeogenesis will be briefly discussed.

The regulation of liver pyruvate kinase by phosphorylation–dephosphorylation has been reviewed previously [26,27].

2. Effect of phosphorylation on the kinetic properties of the enzyme

The main effect of phosphorylation of the rat and pig enzymes is to increase the apparent K_m for PEP in the absence of Fru-1,6-P_2 from 0.3 mM to 0.8–0.9 mM at pH 7.3–7.5 without changing the V_{max} at a saturating concentration of PEP [28,29] (Figure 1). In addition, the Hill coefficient is increased.

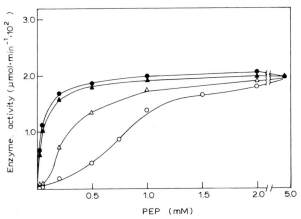

Figure 1. Dependence of rat liver pyruvate kinase activity on PEP concentration. Open and filled symbols represent enzyme activity in the absence and presence of Fru-1,6-P_2, respectively. When Fru-1,6-P_2 was used the concentration was 5 μM. $\triangle$——$\triangle$, $\blacktriangle$——$\blacktriangle$, unphosphorylated pyruvate kinase; $\bigcirc$——$\bigcirc$, $\bullet$——$\bullet$, phosphorylated pyruvate kinase. From Ekman et al. [28].

14

The effect of phosphorylation is most pronounced at a neutral or slightly alkaline pH. At pH 6.3–6.5 there is no difference between the phosphorylated and unphosphorylated enzyme forms when assayed with 0.2 mM PEP. The inhibition of the enzyme by phosphorylation is also counteracted by 5 μM Fru-1,6-P_2, as seen in Figure 1. In the presence of this activator both enzyme forms follow Michaelis–Menten kinetics, with an apparent K_m for PEP of 0.04 mM.

ATP and alanine inhibit phosphorylated pyruvate kinase more strongly than unphosphorylated enzyme [28,29]. On the other hand, Fru-1,6-P_2 activates unphosphorylated enzyme to a greater degree than the phosphorylated form (Figure 2). At 0.2 mM PEP and 1 mM ADP, and in the presence of 1.5 mM ATP and 0.5 mM alanine, rat liver pyruvate kinase is nearly inactive in the absence of Fru-1,6-P_2, as shown in Figure 2. Higher concentrations of Fru-1,6-P_2 are necessary to activate the phosphorylated than the unphosphorylated enzyme. At 25 μM Fru-1,6-P_2 phosphorylation still has an effect on the activity of the enzyme. Under glyconeogenetic conditions the concentration of *free* Fru-1,6-P_2 is probably very low [9]. Thus, the activity of liver pyruvate kinase in the presence of substrates and effectors in the physiological concentration range seems to be decreased by phosphorylation.

Rapid inactivation of pyruvate kinase (assayed at a suboptimal PEP concentration) occurs on incubation of crude extracts from rat hepatocytes with cAMP and ATP [30,31]. This is also the case with a supernatant fraction from human liver [32]. The kinetic properties of these unpurified enzyme

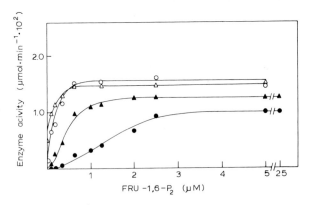

Figure 2. Activity of rat liver pyruvate kinase as a function of Fru-1,6-P_2 concentration in the presence (filled symbols) and absence (open symbols) of ATP and alanine. 0.2 mM PEP was used. When ATP and alanine were present their concentrations were 1.5 and 0.5 mM, respectively. △——△, ▲——▲, unphosphorylated enzyme; ○——○, ●——●, phosphorylated enzyme. From Ekman et al. [28].

preparations are changed in a way very similar to the alteration of purified rat and pig liver pyruvate kinase on phosphorylation [28,29].

3. Effect of phosphorylation on the sensitivity of pyruvate kinase to proteolytic modification with subtilisin

When ^{32}P-labelled liver pyruvate kinase is incubated with a small amount of subtilisin there is a rapid release of [^{32}P]phosphopeptide material without inactivation of the pyruvate kinase activity as measured under optimal conditions [33]. The subunit molecular weight of the modified enzyme is the same, within experimental errors, as that of the subunits of the unmodified enzyme as estimated on polyacrylamide gel electrophoresis in detergent under dissociating conditions. This shows that the phosphorylated site is located near either end of the polypeptide chains.

The phosphate-accepting site may also be cleaved off from unphosphorylated rat and pig enzyme [34] (Figure 3). In this case, however, about ten times more subtilisin is needed to obtain the same rate of modification as that for phosphorylated enzyme (Figure 3). Thus, phosphory-

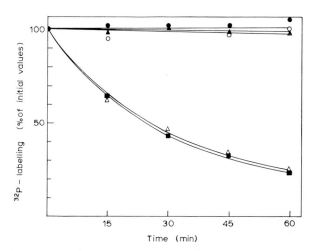

Figure 3. Modification of rat liver pyruvate kinase with subtilisin. The graph shows the time course of release of ^{32}P-labelled phosphopeptides from phosphorylated pyruvate kinase (E-P) and phosphate-accepting sites from unphosphorylated pyruvate kinase (E). ●——●, E; ○——○, E-P incubated without subtilisin; ▲——▲, E incubated with 0.2 μg/ml of subtilisin; ■——■, E incubated with 2 μg/ml and △——△, E-P incubated with 0.2 μg/ml of subtilisin. From Bergström et al. [34].

16

lation of pyruvate kinase makes it more sensitive to proteolytic modification. Since the susceptibility of proteins to proteolytic enzymes in vitro resembles that found in vivo [35], it is possible that phosphorylation of pyruvate kinase in vivo not only affects its activity but also increases its rate of degradation [34].

Subtilisin-modified pyruvate kinase has a still lower affinity for PEP than the phosphorylated enzyme, as shown in Figure 4. V_{max} is the same, however, for the modified and unmodified enzyme forms. In addition, the inhibition of the subtilisin-treated enzyme by ATP is more pronounced than that of unmodified unphosphorylated or phosphorylated pyruvate kinase. Fru-1,6-P$_2$ activates modified pyruvate kinase to a lower degree than the other two forms of the enzyme [34]. Thus, if proteolytically modified pyruvate kinase should exist in vivo as an intermediate in its further degradation, it would be still less active than the phosphorylated form of the enzyme. This would seem reasonable from a physiological point of view.

So far, proteolytically modified liver pyruvate kinase has not been shown to exist in vivo, but certain findings suggest that it might be present in livers from fasted animals. Thus, Kohl and Cottam have purified pyruvate kinase both from fed rats and from animals starved for 72 h [36]. The specific activities of the two enzyme preparations were 220 and 80 units/mg, respectively. Since the enzyme activity was assayed at saturating concentrations of PEP and Fru-

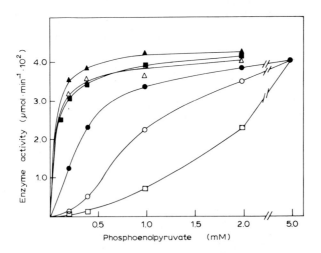

Figure 4. Activity of rat liver pyruvate kinase at different PEP concentrations. When present, the concentration of Fru-1,6-P$_2$ was 20 μM. ●——●, E; ▲——▲, E + Fru-1,6-P$_2$; □——□, subtilisin-modified pyruvate kinase; ■——■, modified enzyme + Fru-1,6-P$_2$; ○——○, E-P; △——△, E-P + Fru-1,6-P$_2$. From Bergström et al. [34].

1,6-P$_2$, the difference should not have been due exclusively to phosphorylation of the enzyme from the starved animal.

Further, Van Berkel et al. have demonstrated that a difference in the apparent affinity of pyruvate kinase for PEP from fed liver versus starved liver persists after incubation of the liver supernatants under optimal conditions for protein kinase activity in vitro [37]. For the completely inactivated enzyme from livers of fed and starved rats K_m values for PEP of 2.8 and 4.4 mM, respectively, were obtained. The authors suggest that the effect of starvation might be caused by partial proteolysis of the enzyme or oxidation of essential thiol groups [37].

4. The protein kinase reaction with native pyruvate kinase as substrate

When rat liver cell sap is chromatographed on a DEAE-cellulose column the same fractions that contain cAMP-stimulated histone kinase activity are also capable of phosphorylating pyruvate kinase under the influence of cAMP [38]. The amino acid sequences have been determined around the specific seryl residue which is phosphorylated in the pig [20] and rat liver [21] enzymes (Table 1). The two sequences are nearly identical and are similar to the sequences of other phosphorylatable proteins with two basic and one or two neutral residues preceding the serine which is phosphorylated. The relative rate of phosphorylation of rat liver pyruvate kinase to that of the β-subunit of rabbit muscle phosphorylase kinase is 35:100 [25].

The protein kinase reaction has been studied with native pyruvate kinase as substrate, using the pig liver enzymes [39]. The phosphorylation rate increases almost linearly with a pyruvate kinase concentration of up to 8 μM subunit

TABLE 1

Amino acid sequences at the phosphorylated site of pyruvate kinase

Enzyme	Amino acid sequence	Reference
Pig liver	Leu-Arg-Arg-Ala-<u>SerP</u>-Leu	[20]
Rat liver	Asx-Thr-Lys-Gly-Pro-Glx-Ile-Glx- Thr-Gly-Val-Leu-Arg-Arg-Ala-<u>SerP</u>- Val-Ala-Glx-Leu	[21]

<u>SerP</u>, phosphoserine

(the highest concentration used), indicating a rather high apparent K_m of the protein kinase for pyruvate kinase. In comparison with histone, pyruvate kinase is a better substrate at high pH values (up to pH 9). Thus, the form of the enzyme which has cooperative kinetics with respect to PEP and which dominates at higher pH values is phosphorylated more rapidly than the form with Michaelis–Menten kinetics which exists at a lower pH. The allosteric inhibitors alanine and phenylalanine also increase the rate of phosphorylation, but the activator Fru-1,6-P$_2$ has no effect under conditions used [39].

Felíu et al. have studied the cAMP-stimulated and Mg-ATP-requiring inactivation of pyruvate kinase in a crude extract of isolated rat hepatocytes [30]. They found that the inactivation was counteracted by physiological concentrations of PEP or Fru-1,6-P$_2$. This effect of the two metabolites on the phosphorylation of pyruvate kinase has also been described recently by Pilkis et al. [31].

To summarize, liver pyruvate kinase seems to be a better substrate for cAMP-stimulated protein kinase when it is in a less active conformation, and vice versa. This seems to be appropriate, since an allosteric inhibition will then lead to a further inhibition by phosphorylation. The activators PEP and Fru-1,6-P$_2$ also keep the enzyme active by counteracting a phosphorylation.

5. Protein kinase-catalyzed phosphorylation of synthetic peptides with amino acid sequences representing the phosphate-accepting site of pyruvate kinase

Alkali-inactivated pig liver pyruvate kinase has been found to be more rapidly phosphorylated by ATP and cAMP-stimulated protein kinase than the native enzyme [22]. This is even more so for a cyanogen bromide fragment of the enzyme. The finding that peptides may be phosphorylated by cAMP-stimulated protein kinase was also reported at about the same time by other authors [40–43]. This indicated that the local primary structure of a small part of a phosphorylatable protein fulfils the minimal structural requirements for phosphorylation.

Since the amino acid sequence near the phosphate-accepting seryl residue in rat liver pyruvate kinase is known (Table 1), it was possible to synthesize a number of peptides with sequences corresponding to different lengths near this seryl residue in order to investigate the minimal structural requirements for phosphorylation of pyruvate kinase [44]. The smallest peptide which is phosphorylated at a rate similar to that for the native enzyme is the pentapeptide Arg-Arg-Ala-Ser-Val (Table 2). Both arginyl residues are necessary for phosphorylation at an appreciable rate, since removal of arginine decreases the phosphorylation rate 50-fold. If both arginyl residues are removed or a leucyl residue is substituted for either arginyl residue, no measurable phosphorylation takes place. Lengthening of the peptide increases its rate of phosphorylation, as seen in Table 2. In addition, the affinity of the protein kinase for the peptide substrates increases, the K_m values being 0.08 mM and less than 0.01 mM for the pentapeptide and the heptapeptide Leu-Arg-Arg-Ala-Ser-Val-Ala, respectively. This means that the K_m of the heptapeptide is not too far from the concentration of pyruvate kinase in rat liver [1,18,19]. When native pyruvate kinase is phosphorylated at a subunit concentration of 0.01 mM the rate is about the same as for the active pentapeptide at the same concentration [44].

Kemp et al. have studied the phosphorylation of synthetic peptides related to the phosphorylated site of pig liver pyruvate kinase [45]. Their results are in general agreement with ours [44]. Of special importance is the finding that the presence of two arginyl residues in the peptides, instead of one, drastically decreases the apparent K_m of the protein kinase for the substrate peptides [45]. The kinetics of the phosphorylation of the peptide Leu-Arg-Arg-Ala-Ser-Leu-Gly with cAMP-stimulated protein kinase from calf thymus have been studied by Pomerantz et al. [46]. They found a V_{max} of 50.6 μmol/min/mg of kinase and an apparent K_m of 63 μM. These figures are similar to those

TABLE 2

Sequence of synthetic peptides and relative rate of their phosphorylation by [^{32}P]ATP and the catalytic subunit of a cAMP-stimulated protein kinase from rat liver

Peptide	Relative rate of phosphorylation
Leu-Arg-Arg-Ala-Ser-Val-Ala	207
Arg-Arg-Ala-Ser-Val-Ala	193
Arg-Ala-Ser-Val-Ala	2
Ala-Ser-Val-Ala	<1
Arg-Arg-Ala-Ser-Val	100
Arg-Arg-Ala-Ser	<1
Leu-Arg-Ala-Ser-Val	<1
Arg-Leu-Ala-Ser-Val	<1
Arg-Arg-Ala-Ser-Gly	7
Arg-Arg-Ala-Ser-Phe	138
Arg-Arg-Ala-Ser-Lys	17
Arg-Arg-Ala-Thr-Val	<1

From Zetterqvist et al. [44].

obtained under similar conditions by Kemp et al. [45] with use of the catalytic subunit of the bovine skeletal muscle enzyme, namely 20 μmol/min/mg and 16 μM, respectively.

6. Dephosphorylation of pyruvate kinase with phosphoprotein phosphatase

Maximally phosphorylated rat liver pyruvate kinase may be almost completely reactivated by incubation with a histone phosphatase preparation from rat liver call sap [38]. After dephosphorylation the curve of pyruvate kinase activity versus PEP concentration is almost the same as before phosphorylation. The total amount of phosphatase activity in rat liver cytosol has been estimated to be sufficient to completely dephosphorylate all the pyruvate kinase present in normally fed rats in about 10 s [47].

On chromatography of rat liver cell sap on a DEAE-cellulose column three peaks of phosphoprotein phosphatase activity are obtained, using phosphorylated protamine, histone or rabbit muscle phosphorylase a as substrate (Figure 5). Only the last two peaks (B and C in Figure 5) are active on phosphopyruvate kinase. The molecular weights of these enzyme fractions are approximately 250 000 and 140 000, respectively, as determined by chromatography on a calibrated Sephadex G-200 column [47]. The last fraction (peak C) is proportionately more active on phosphopyruvate kinase than on phosphoprotamine in comparison with peak B. Phosphatase C is more labile than phosphatase B and, regardless of the substrate used, is partially or completely inactivated by the ethanol treatment used earlier during purification of phosphoprotein phosphatase [48]. On the other hand, the action of fraction B on phosphorylase a is increased by 50−100% after ethanol treatment but its action on phosphopyruvate kinase is decreased to about 70% of the initial activity.

Phosphatase B is strongly activated by magnesium and manganese ions, but the increase of the activity of fraction C is less than 2-fold [47]. Both fractions are moderately inhibited by EDTA, orthophosphate, fluoride and ATP. The inhibition of 1 mM ATP is overcome by the addition of 2.5 mM $MgCl_2$ [47].

A highly purified phosphatase from rat liver has been obtained (Titanji, V.P.K., Zetterqvist, Ö. and Engström, L. [1979], in preparation). This enzyme is active on phosphorylase a, phosphoprotamine, phosphohistone and phosphopyruvate kinase, approximately at the same rate. It is not inhibited by the phosphorylase phosphatase inhibitors from rabbit muscle. Its molecular weight is 32 000−35 000. Its relation to the phosphatases B and C described

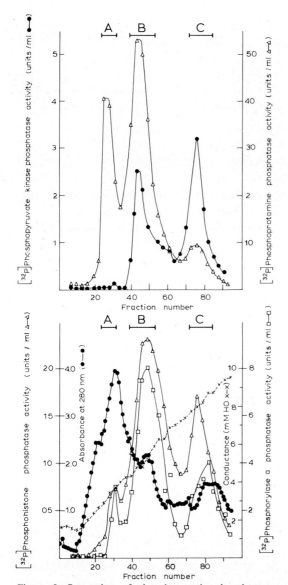

Figure 5. Separation of phosphoprotein phosphatase on DEAE-cellulose. Rat livers were homogenized with 3 vol of 250 mM sucrose, containing 15 mM β-mercaptoethanol, 1 mM EDTA (pH 7.0) and 0.1 mM phenylmethylsulfonyl fluoride. The postmicrosomal supernatant was applied to a DEAE-cellulose column (Whatman DE-52) in equilibrium with 40 mM NaCl in a buffer solution containing 10 mM imidazole/HCl (pH 7.5), 15 mM mercaptoethanol and 2.5 mM MgCl$_2$. The enzymes were eluted with a linear gradient of 40 to 350 mM NaCl in the same buffer solution. The protein concentration of the fractions was estimated as absorbance at 280 nm. In assays of the phosphoprotein phosphatase activity, the substrates were: upper panel [^{32}P]-phosphopyruvate kinase ($\bullet$——$\bullet$), [^{32}P]phosphoprotamine ($\triangle$——$\triangle$); lower panel [^{32}P]-phosphohistone ($\triangle$——$\triangle$), and [^{32}P]phosphorylase a ($\square$——$\square$). From Titanji [47].

above is not clear. Since ethanol treatment is used during the purification procedure, it seems possible that it represents a catalytic subunit of phosphatase B and not C, which is labile against ethanol [47].

The apparent K_m for dephosphorylation of phosphopyruvate kinase with the highly purified enzyme is 27 μM (expressed as phosphate content) [48]. This is about one order of magnitude higher than the pyruvate kinase subunit concentration in normally fed rats. The enzyme reaction is slightly activated by divalent ions but does not seem to be influenced by approximately physiological concentrations of Fru-1,6-P$_2$, PEP, ADP or ATP in the presence of divalent ions [48].

Thus, rat liver cell sap contains a high phosphoprotein phosphatase activity towards phosphopyruvate kinase. Under mild conditions the activity can be separated into two high-molecular-weight fractions. It is not known which of the fractions is most important in vivo or from which fraction the low-molecular-weight phosphatase is obtained. Also, very little is known about the regulation of the phosphatase activity acting on pyruvate kinase.

7. Dephosphorylation with phosphoprotein phosphatase of phosphopeptides corresponding to the phosphorylated site of pyruvate kinase

It has recently been shown that synthetic phosphopeptides are substrates of a purified phosphoprotein phosphatase from rat liver [49]. In Table 3 K_m and

TABLE 3

[^{32}P]Phosphopeptide substrates of rat-liver phosphoprotein phosphatase

Peptide	Apparent K_m (mM)	Relative V_{max}
Gly-Val-Leu-Arg-Arg-Ala-SerP-Val-Ala	0.24	1.0
Val-Leu-Arg-Arg-Ala-SerP-Val-Ala	0.31	1.0
Leu-Arg-Arg-Ala-SerP-Val-Ala	0.57	1.0
Arg-Arg-Ala-SerP-Val-Ala	0.28	1.0
Arg-Ala-SerP-Val-Ala	0.41	1.0
Ala-SerP-Val-Ala[a]	0.07	0.4
Arg-Arg-Ala-SerP-Val[b]	–	–
Leu-Arg-Arg-Ala-SerP-Val[b]	–	–

From Titanji et al. [49].
[a] From Titanji, V.P.K., Ragnarsson, U., Humble, E., and Zetterqvist, Ö. (To be published).
[b] No detectable dephosphorylation.

relative V_{max} values are given for a number of phosphopeptides with amino acid sequences from the phosphorylated site of rat liver pyruvate kinase. V_{max} for the largest peptides are about the same as V_{max} for phosphoprotamine under saturation conditions [49]. At 20 μM phosphopeptide the dephosphorylation rate is about five times lower than that for native phosphopyruvate kinase, 20 μM, with respect to the phosphate moiety of the enzyme.

The shortest peptides dephosphorylated at an appreciable rate is the tetrapeptide Ala-SerP-Val-Ala. Thus, none of the two arginyl residues which are necessary for the corresponding protein kinase reaction are needed for the phosphatase reaction. At least one amino acid residue preceding and two residues after the phosphorylserine are required for the peptide to be a substrate. However, even the lowest values for K_m are at least one order of magnitude higher than the intracellular concentration of pyruvate kinase subunits. This suggests the possibility that additional specificity determinants reside on the pyruvate kinase molecule besides the local primary structure of the phosphorylated site.

8. Rapid hormonal regulation of pyruvate kinase activity in liver cells in vitro and in vivo

As discussed in previous review articles [26,27], several research groups working on whole animals [14], perfused liver [50] or isolated hepatocytes [51–54] have demonstrated an inhibition of pyruvate kinase at suboptimal substrate concentrations after treatment with glucagon or cAMP. The hormone induces a rapid increase in the apparent K_m for PEP without changing the V_{max} of the enzyme [50–52,54]. In addition, glucagon causes an increase in the Hill coefficient and in the affinity of the enzyme for the inhibitors ATP and alanine [50,51]. The affinity of pyruvate kinase for the activator Fru-1,6-P_2 in isolated hepatocytes decreases after treatment of the cells with glucagon [52]. In several cases the effects of glucagon have been shown to be counteracted or reversed by insulin [14,50–52]. The effects of glucagon on the pyruvate kinase activity in intact cells are very similar to those observed on phosphorylation of the purified enzyme by cAMP-stimulated protein kinase [28,29].

These results have recently been verified and extended [37,55–59]. When hepatocytes are incubated with 10 mM alanine the inhibition of pyruvate kinase by glucagon is enhanced, lowering the hormone concentration needed for a half-maximal effect from 0.3 to 0.1 nM [59]. This shows that the

24

phosphorylation of the enzyme is facilitated by an allosteric inhibitor also in intact cells. On the other hand, a high intracellular concentration of PEP seems to inhibit the inactivation of liver pyruvate kinase caused by glucagon treatment in vivo [30].

9. *Phosphorylation of pyruvate kinase in intact cells*

In some investigations the phosphorylation of pyruvate kinase in intact cells has been studied during treatment with glucagon. When rat liver slices are incubated in a medium containing [^{32}P]orthophosphate, pyruvate kinase isolated by immunoadsorbent chromatography is partially phosphorylated in the absence of added hormone (0.2 mol of phosphate/mol of enzyme subunit) [60]. Upon incubation with 10^{-7}M glucagon, the phosphorylation increases within 2 min to 0.6–0.7 mol/mol of enzyme subunit (Figure 6). At the same time the enzyme activity at a suboptimal PEP concentration is inhibited and the content of cAMP is increased. Figure 7 shows the relationship between degree of phosphorylation and enzyme activity at different hormone concentrations. Since the enzyme is inactivated in parallel to its phosphorylation it

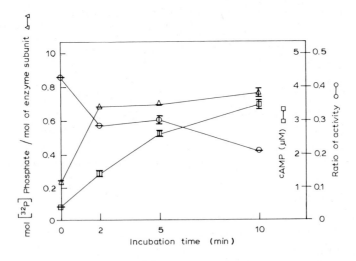

Figure 6. Velocity of pyruvate kinase phosphorylation in rat liver slices incubated for different times with 10^{-7} M glucagon after preincubation in a medium containing [^{32}P]orthophosphate. Samples were analyzed for phosphorylation of pyruvate kinase (△——△), ratio of enzyme activity at 0.5 to that at 5 mM PEP (○——○) and cAMP concentration (□——□). From Ljungström and Ekman [60].

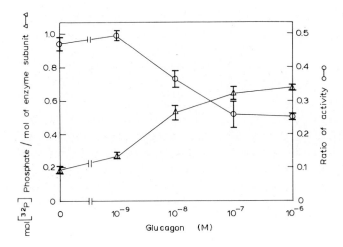

Figure 7. Dependence of phosphorylation of pyruvate kinase on glucagon concentration. ^{32}P-labelled rat liver slices were incubated for 10 min with different concentrations of glucagon. Samples were analyzed for phosphorylation (△——△) and ratio of enzyme activity at 0.5 to that at 5 mM PEP (○——○). From Ljungström and Ekman [60].

may be concluded that the inactivation is caused by the phosphate incorporation. The same [^{32}P]phosphopeptide pattern is obtained from pyruvate kinase isolated from ^{32}P-labelled slices and from purified pyruvate kinase incubated in vitro with cAMP-stimulated protein kinase and with [^{32}P]ATP, respectively [60]. This strongly indicates that the same seryl residue in the enzyme is phosphorylated in vivo as in vitro.

Riou et al. injected rats with [^{32}P]orthophosphate intraperitoneally and rapidly purified liver pyruvate kinase to homogeneity in the presence of potassium fluoride and EDTA in order to inhibit the phosphoprotein phosphatase and protein kinase reactions, respectively [61]. They also isolated the enzyme by specific immunoprecipitation of partially purified preparations. In both cases the enzyme was labelled with [32]P. Glucagon administration in vivo increased the ^{32}P-incorporation 200–300%, an increase which was correlated with an inhibition of the enzyme activity and an elevation of the levels of cAMP in the liver.

Ishibashi and Cottam have demonstrated that the addition of 10^{-6}M glucagon or 10^{-3}M cAMP to [^{32}P]orthophosphate-incubated hepatocytes results in a 3-fold increase in the incorporation of ^{32}P into L-type pyruvate kinase and reduced enzyme activity [62,63].

To conclude, it seems to be fairly well ascertained that the glucagon-induced changes of the catalytic properties of liver pyruvate kinase are due to phosphorylation of the enzyme.

10. Relationship of mammalian liver L-type pyruvate kinase to other pyruvate kinases

In mammals, L-type pyruvate kinase is present in renal cortex as a minor component [64]. This enzyme has been purified from rat kidneys by Berglund and found to be phosphorylated by cAMP-stimulated protein kinase [65]. The enzyme is immunologically identical with the rat liver enzyme. The amino acid sequences around the seryl residue phosphorylated seem to be same in the two enzymes, as shown by an electrophoretic comparison of the [^{32}P]phospho-peptides obtained from ^{32}P-labelled enzyme from the two tissues [65]. The affinity for PEP is decreased by phosphorylation of the L-type kidney pyruvate kinase. This effect is reversed by incubation of the phosphorylated enzyme with phosphoprotein phosphatase [65].

Pyruvate kinase from erythrocytes is of the L-type, as shown by immunological methods [66]. However, the rat erythrocyte enzyme is not phosphorylated by cAMP-stimulated protein kinase and [^{32}P]ATP [67]. Its kinetic properties differ from those of the rat liver enzyme. But after proteolytic modification of rat liver pyruvate kinase with a small amount of subtilisin, which removes the phosphorylated site, the catalytic properties of the liver enzyme are very similar to those of the erythrocyte enzyme. It has therefore been suggested that the latter enzyme might represent a modified liver type enzyme [67]. On the other hand, indications have been obtained that the subunits of the corresponding enzymes from humans are coded by a single gene, but that the subunits of the liver enzyme seem to be a degradation product of the enzyme in erythroblasts and young red cells [68]. The subunit molecular weights are 58 000 and 63 000, respectively. In mature erythrocytes the enzyme contains two of each type of subunit [68]. No attempts to phosphorylate the human erythrocyte pyruvate kinase have been reported. It therefore seems rather difficult at present to evaluate the relationships between the liver and erythrocyte enzymes.

Chicken liver has been reported to contain both type L and type A pyruvate kinase [69]. Both isozymes are reported to be phosphorylated by a cAMP-independent protein kinase [70]. The M_2-type enzyme which predominates in the chicken liver is completely inactivated by the incorporation of 4 mol of phosphate/mol of tetrameric enzyme. The protein kinase involved is bound firmly to the pyruvate kinases. It phosphorylates casein and phosvitin but not histone. GTP in addition to ATP can function as phosphate donor [70]. The phosphorylation is reversed by the presence of a phosphoprotein phosphatase [71].

With regard to the mammalian pyruvate kinase isozymes no cAMP-independent phosphorylation has been described so far.

11. Discussion

Using the criteria of Krebs [24] and of Nimmo and Cohen [25], it is clear that liver pyruvate kinase, at least in rats, belongs to the group of enzymes whose activities are regulated through a reversible phosphorylation under the influence of cAMP. Rat liver contains cAMP-stimulated protein kinase and phosphoprotein phosphatase, which have a strong action on native pyruvate kinase. The enzyme activity is, as expected, decreased on phosphorylation in the presence of substrates and effectors in the physiological concentration ranges. Furthermore, phosphorylation might increase the degradation rate of the enzyme in vivo, since phosphorylated pyruvate kinase is more sensitive to proteolytic modification than the unphosphorylated enzyme [34].

Glucagon, cAMP and starvation induce changes in the kinetic properties of pyruvate kinase in vivo and in intact cells which are very similar to those caused by phosphorylation of the enzyme in vitro. The effects are reversed or counteracted by insulin [14, 50–52]. In the case of isolated hepatocytes, the glucagon-induced inactivation of pyruvate kinase is also reversible in the absence of insulin, but at a fairly slow rate [51]. The inactivation of the enzyme caused by glucagon or cAMP occurs roughly in parallel with an increased phosphorylation, as demonstrated on whole animals [61], rat liver slices [60] and isolated hepatocytes [62,63]. In the case of liver slices strong evidence has been produced that the same seryl residue in pyruvate kinase is phosphorylated in intact cells as on incubation of purified enzyme with protein kinase and [^{32}P]ATP [60].

Glucagon and cAMP stimulate gluconeogenesis [72]. The exact role of the glucagon-induced phosphorylation of pyruvate kinase in the regulation of gluconeogenesis under different conditions is difficult to decide at present. Several reports indicate that pyruvate kinase is one site for hormonal regulation of gluconeogenesis [50,51,55,73]. In studies on isolated hepatocytes incubated with glucagon, cAMP or epinephrine in the presence or absence of insulin, Felíu et al. found inverse effects on the activity of pyruvate kinase and on gluconeogenesis [51]. The authors concluded that the large variations in the enzyme activity are at least partly responsible for the changes in the rate of gluconeogenesis.

Rognstad and Katz have shown that glucagon and epinephrine stimulate gluconeogenesis from 20 mM lactate with about three times greater effects on hepatocytes from fed rats than on those from fasted rats [55]. Neither of the hormones had any effect on the pyruvate kinase flux in cells from starved rats, indicating that the effect on gluconeogenesis is exerted primarily at other sites than that on pyruvate kinase. In cells from fed rats both hormones stimulated gluconeogenesis considerably, but only glucagon caused a marked reduction on the pyruvate kinase flux. This shows that there is a difference in action between glucagon and epinephrine [55]. At lower glucagon concentrations the effect on gluconeogenesis is proportionately greater than that on the pyruvate kinase flux. It is therefore concluded that glucagon acts on at least one more site in its effect on gluconeogenesis than in its effect on pyruvate kinase flux [55].

Foster and Blair have demonstrated a good correlation between inhibition of pyruvate kinase and lactate formation from 10 mM dihydroxyacetone in hepatocytes from fasted rats [59]. However, when 10 mM fructose was used as substrate 1 μM glucagon caused no marked inhibition of lactate formation even though this hormone partially inactivated pyruvate kinase when tested at a low PEP concentration [59]. The reason for this unchanged flux through the pyruvate kinase reaction is not clear. However, it must be kept in mind that the flux in question in this metabolic step must also be dependent on the concentrations of free PEP as well as of Fru-1,6-P_2 and other effectors of the enzyme.

Epinephrine decreases pyruvate kinase activity in whole animals [16] and in isolated hepatocytes [51,59,74,75]. The mechanism of action of epinephrine on the pyruvate kinase reaction and on gluconeogenesis is not known. The hormone probably inhibits pyruvate kinase by an α-receptor-mediated, cAMP-independent mechanism [74]. Certain findings have also indicated that both α- and β-receptor mechanisms may be involved [75].

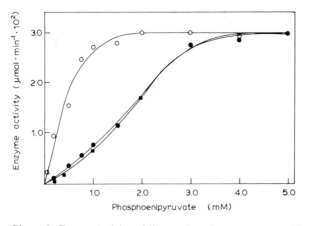

Figure 8. Enzyme activity of liver and erythrocyte pyruvate kinase at different PEP concentrations. ○——○, liver enzyme; ●——●, erythrocyte enzyme and ■——■, subtilisin-modified liver enzyme. From Dahlqvist-Edberg [67].

In conclusion, it is fairly well established that liver pyruvate kinase type L is an enzyme whose activity is regulated by a reversible phosphorylation under the influence of cAMP. The exact role of this phosphorylation in the regulation of carbohydrate metabolism in vivo is not known. However, strong evidence has been accumulated that phosphorylation of pyruvate kinase is one important way of regulation of glycolysis and gluconeogenesis.

In addition, synthetic peptides with amino acid sequences related to the phosphorylated site of liver pyruvate kinases have been a useful tool in studies of the mechanism of action of cAMP-stimulated protein kinase [44–46] and phosphoprotein phosphatases [49].

References

1. Scrutton, M.C. and Utter, M.F. (1968) Annu. Rev. Biochem., 37, 249–302.
2. Crisp, D.M. and Pogson, C.I. (1972) Biochem. J., 126, 1009–1023.
3. Van Berkel, T.J.C., Koster, J.F. and Hülsmann, W.C. (1972) Biochim. Biophys. Acta, 276, 425–429.
4. Krebs, H.A. and Eggleston, L.V. (1965) Biochem. J., 94, 3c–4c.
5. Seubert, W. and Schoner, W. (1971) Curr. Top. Cell. Regul., 3, 237–267.
6. Tanaka, T., Harano, Y., Sue, F. and Morimura, H. (1967) J. Biochem. (Tokyo), 62, 71–91.
7. Carminatti, H., Jiménez de Asúa, L., Recondo, E., Passeron, S. and Rozengurt, E. (1968) J. Biol. Chem., 243, 3051–3056.
8. Flory, W., Peczon, B.D., Koeppe, R.E. and Spivey, H.O. (1974) Biochem. J., 141, 127–131.
9. Van Berkel, T.J.C., Koster, J.F., Kruyt, J.K. and Hülsmann, W.C. (1974) Biochim. Biophys. Acta, 370, 450–458.
10. Exton, J.H. and Park, C.R. (1968) J. Biol. Chem., 243, 4189–4196.
11. Exton, J.H. and Park, C.R. (1969) J. Biol. Chem., 244, 1424–1433.
12. Sutherland, E.W. (1972) Science, 177, 401–408.
13. Kuo, J.F. and Greengard, P. (1969) J. Biol. Chem., 244, 3417–3419.
14. Taunton, O.D., Stifel, F.B., Greene, H.L. and Herman, R.H. (1972) Biochem. Biophys. Res. Commun., 48, 1663–1670.
15. Taunton, O.D., Stifel, F.B. Greene, H.L. and Herman, R.H. (1974) J. Biol. Chem., 249, 7228–7239.
16. Stifel, F.B., Taunton, O.D., Greene, H.L. and Herman, R.H. (1974) J. Biol. Chem., 249, 7240–7244.
17. Greene, H.L., Taunton, O.D., Stifel, F.B. and Herman, R.H. (1974) J. Clin. Invest., 53, 44–51.
18. Ljungström, O., Hjelmquist, G. and Engström, L. (1974) Biochim. Biophys. Acta, 358, 289–298.
19. Engström, L., Berglund, L., Bergström, G., Hjelmquist, G. and Ljungström, O. (1974) in Lipmann Symposium: Energy, Regulation and Biosynthesis in Molecular Biology (Richter, D., ed.), pp. 192–204, de Gruyter, Berlin.
20. Hjelmquist, G., Andersson, J., Edlund, B. and Engström, L. (1974) Biochem. Biophys. Res. Commun., 61, 559–563.
21. Edlund, B., Anderson, J., Titanji, V., Dahlqvist, U., Ekman, P., Zetterqvist, Ö. and Engström, L. (1975) Biochem. Biophys. Res. Commun., 67, 1516–1521.

30

22. Humble, E., Berglund, L., Titanji, V., Ljungström, O., Edlund, B., Zetterqvist, Ö. and Engström, L. (1975) Biochem. Biophys. Res. Commun., 66, 614–621.
23. Cohen, P., Watson, D.C. and Dixon, G.H. (1975) Eur. J. Biochem., 51, 79–92.
24. Krebs, E.G. (1973) in Endocrinology, Proceedings of the Fourth International Congress Endocrinology., pp. 17–29, Excerpta Medica, Amsterdam.
25. Nimmo, H.G. and Cohen, P. (1977) in Advances in Cyclic Nucleotide Research, Vol. 8, pp. 145–266, Raven, New York.
26. Engström, L. (1978) in Current Topics in Cellular Regulation, Vol. 13, pp. 29–51, Academic Press, New York.
27. Engström, L. (1978) in FEBS 11th Meeting Copenhagen 1977, Symposium A 1: Regulatory Mechanisms of Carbohydrate Metabolism, Vol. 42, pp. 53–60, Pergamon, Oxford.
28. Ekman, P., Dahlqvist, U., Humble, E. and Engström, L. (1976) Biochim. Biophys. Acta, 429, 374–382.
29. Ljungström, O., Berglund, L. and Engström, L. (1976) Eur. J. Biochem., 68, 497–506.
30. Felíu, J.E., Hue, L. and Hers, H.-G. (1977) Eur. J. Biochem., 81, 609–617.
31. Pilkis, S.J., Pilkis, J. and Claus, T.H. (1978) Biochem. Biophys. Res. Commun., 81, 139–146.
32. Van den Berg, G.B., Van Berkel, T.J.C. and Koster, J.F. (1978) Biochem. Biophys. Res. Commun., 82, 859–864.
33. Bergström, G., Ekman, P., Dahlqvist, U., Humble, E. and Engström, L. (1975) FEBS Lett., 56, 288–291.
34. Bergström, G., Ekman, P., Humble, E. and Engström, L. (1978) Biochim. Biophys. Acta, 532, 259–267.
35. Goldberg, A.L. and Dice, J.F. (1974) Ann. Rev. Biochem., 43, 835–869.
36. Kohl, E.A. and Cottam, G.L. (1977) Biochim. Biophys. Acta, 484, 49–58.
37. Van Berkel, T.J.C., Kruut, J.K. Van den Berg, G.B. and Koster, J.F. (1978) Eur. J. Biochem., 92, 553–561.
38. Titanji, V.P.K., Zetterqvist, Ö. and Engström, L. (1976) Biochim. Biophys. Acta, 422, 98–108.
39. Berglund, L., Ljungström, O. and Engström, L. (1977) J. Biol. Chem., 252, 613–619.
40. Daile, P. and Carnegie, P.R. (1974) Biochem. Biophys. Res. Commun., 61, 852–858.
41. Daile, P., Carnegie, P.R. and Young, J.D. (1975) Nature, 257, 416–418.
42. Kemp, B.E., Bylund, D.B., Huang, T.-S. and Krebs, E. (1975) Proc. Natl. Acad. Sci. USA, 72, 3448–3452.
43. Kemp, B.E., Benjamini, E. and Krebs, E.G. (1976) Proc. Natl. Acad. Sci. USA, 73, 1038–1042.
44. Zetterqvist, Ö., Ragnarsson, U., Humble, E., Berglund, L. and Engström, L. (1976) Biochem. Biophys. Res. Commun., 70, 696–703.
45. Kemp, B.E., Graves, D.J., Benjamini, E. and Krebs, E.G. (1977) J. Biol. Chem., 252, 4888–4894.
46. Pomerantz, A.H., Allfrey, V.G., Merrifield, R.B. and Johnson, E.M. (1977) Proc. Natl. Acad. Sci. USA, 74, 4261–4265.
47. Titanji, V.P.K. (1978) Uppsala, J. Med. Sci., 83, 129–134.
48. Titanji, V.P.K. (1977) Biochim. Biophys. Acta, 481, 140–151.
49. Titanji, V.P.K., Zetterqvist, Ö and Ragnarsson, U. (1977) FEBS Lett. 78, 86–90.
50 Blair, J.B., Cimbala, M.A., Foster, J.L. and Morgan, R.A. (1976) J. Biol. Chem., 251, 3756–3762.
51. Felíu, J.E., Hue, L. and Hers, H.-G. (1976) Proc. Natl. Acad. Sci. USA, 73, 2762–2766.
52. Van Berkel, T.J.C., Kruijt, J.K., Koster, J.F. and Hülsmann, W.C. (1976) Biochem. Biophys. Res. Commun., 72, 917–925.
53. Friedrichs, D. (1975) FEBS Meeting 10th, Abstract No. 1455.
54. Riou, J.P., Claus, T.H. and Pilkis, S.J. (1976) Biochem. Biophys. Res. Commun., 73, 591–599.

55. Rognstad, R. and Katz, J. (1977) J. Biol. Chem., 252, 1831–1833.
56. Van Berkel, T.J.C., Kruijt, J.K. and Koster, J.F. (1977) Biochim. Biophys. Acta, 500, 267–276.
57. Van Berkel, T.J.C., Kruijt, J.K. and Koster, J.F. (1977) Eur. J. Biochem., 81, 423–432.
58. Van Berkel, T.J.C., Kruijt, J.K. and Koster, J.F. (1978) Biochem. Soc. Trans., 6, 147–149.
59. Foster, J.L. and Blair, J.B. (1978) Arch. Biochem. Biophys., 189, 263–276.
60. Ljungström, O. and Ekman, P. (1977) Biochem. Biophys. Res. Commun., 78, 1147–1155.
61. Riou, J.P., Claus, T.H. and Pilkis, S.J. (1978) J. Biol. Chem., 253, 656–659.
62. Ishibashi, H. and Cottam, G.L. (1978) Biochem. Biophys. Res. Commun., 85, 900–905.
63. Ishibashi, H. and Cottam, G.L. (1978) J. Biol. Chem., 253, 8767–8771.
64. Jiménez de Asúa, L., Rozengurt, E. and Carminatti, H. (1971) FEBS Lett., 14, 22–24.
65. Berglund, L. (1978) Biochim. Biophys. Acta, 524, 68–77.
66. Marie, J., Kahn, A. and Boivin, P. (1977) Biochim. Biophys. Acta, 481, 96–104.
67. Dahlqvist-Edberg, U. (1978) FEBS Lett., 88, 139–143.
68. Kahn, A., Marie, J., Garreau, H. and Sprengers, E.D. (1978) Biochim. Biophys. Acta, 523, 59–74.
69. Eigenbrodt, E. and Schoner, W. (1977) Hoppe-Seyler's Z. Physiol. Chem., 358, 1033–1046.
70. Eigenbrodt, E., Abdel-Fattah Mostafa, M. and Schoner, W. (1977) Hoppe-Seyler's Z. Physiol. Chem., 358, 1047–1055.
71. Eigenbrodt, E. and Schoner, W. (1977) Hoppe-Seyler's Z. Physiol. Chem., 358, 1057–1067.
72. Exton, J.H., Mallette, L.E., Jefferson, L.S., Wong, E.H.A., Friedmann, N., Miller, T.B. and Park, C.R. (1970) Recent Prog. Horm. Res., 26, 411–461.
73. Pilkis, S.J., Riou, J.P. and Claus, T.H. (1976) J. Biol. Chem., 251, 7841–7852.
74. Chan, T.M. and Exton, J.H. (1978) J. Biol. Chem., 253, 6393–6400.
75. Kemp, B.E. and Clark, M.G. (1978) J. Biol. Chem., 253, 5147–5154.

CHAPTER 3

The regulation of fatty acid synthesis by reversible phosphorylation of acetyl-CoA carboxylase

D. GRAHAME HARDIE

1. Introduction

Higher organisms require fatty acids for incorporation into membrane phospholipids and as a long-term energy store in the form of triglyceride. This requirement for fatty acids can be met either from dietary intake or by de novo synthesis from acetyl-CoA. Fatty acid synthesis from cytoplasmic acetyl-CoA is catalysed by two enzymes, acetyl-CoA carboxylase and fatty acid synthase. The overall reactions are as follows:

(1) Acetyl-CoA carboxylase
$$\text{acetyl-CoA} + CO_2 + \text{ATP} \rightarrow \text{malonyl-CoA} + \text{ADP} + P_i$$

(2) Fatty acid synthase
$$\text{acetyl-CoA} + 7\,\text{malonyl-CoA} + 14\,\text{NADPH} + 14\,H^+ \rightarrow$$
$$\text{palmitic acid} + 7\,CO_2 + 8\,\text{CoA} + 14\,\text{NADP}^+ + 6\,H_2O$$

From the stoichiometry of these two reactions it can be seen that the synthesis of 1 molecule of palmitic acid involves the hydrolysis of 7 molecules of ATP and the oxidation of 14 molecules of NADPH. Fatty acid synthesis therefore makes large demands on metabolic energy and the finding that the pathway is stringently controlled is not unexpected. Both enzymes are subject to control at the level of protein synthesis and degradation, a topic which has been reviewed elsewhere [1,2]. The possibility that the pathway is also subject to short-term regulation by modulation of enzyme activity has been given little attention until quite recently. However, as discussed in Section 2, it is now clear that several hormones as well as dietary factors can affect fatty acid synthesis with a rapidity which cannot be explained by changes in the levels of enzyme protein. The hypothesis which I wish to develop in this article is that

Cohen (ed.) Recently discovered systems of enzyme regulation by reversible phosphorylation
© Elsevier/North-Holland Biomedical Press, 1980

34

many, if not all, of these effects can be explained by changes in the state of phosphorylation of acetyl-CoA carboxylase.

2. *The hormonal control of fatty acid synthesis*

In Table 1 I have summarised a large body of literature on the short-term regulation of fatty acid synthesis by hormones. These results were obtained either in vivo, with perfused organs, or with isolated cell preparations and I have used two criteria in selecting the data for inclusion in Table 1. Firstly, the

TABLE 1

A summary of selected literature on the short-term regulation of fatty acid synthesis by hormones

Hormone	Experimental system	Effect on fatty acid synthesis	Effect on acetyl-CoA carboxylase activity	Reference
Glucagon	Rat liver in vivo	↓	↓	[3]
	Rat liver in vivo	↓	−	[4]
	Chicken liver cells	↓	−	[5]
	Rat liver cells	↓	−	[4]
	Rat liver cells	↓	↓	[6]
	Rat liver cells	−	↓	[7]
Adrenaline	Perfused mouse liver	↓	↓*	[8]
	Rat adipose tissue in vivo	−	↓*	[9]
	Rat fat pads	−	↓*	[10]
	Rat fat pads	−	↓*	[11]
	Rat fat cells	−	↓*	[11]
Insulin	Rat liver in vivo	↑	↑*	[12]
	Rat liver cells	↑	↑	[6]
	Rat adipose tissue in vivo	↑	↑*	[12]
	Rat brown adipose tissue in vivo	↑	↑*	[13]
	Rat fad pads	−	↑	[14]
	Rat fad pads	−	↑*	[15]
Vasopressin	Perfused mouse liver	↓	↓*	[8]
Angiotensin II	Perfused mouse liver	↓	↓*	[16,17]

↑ Increase
↓ Decrease
− Not determined
* Total amount of acetyl-CoA carboxylase unchanged by hormone treatment

effects must have been sufficiently rapid to render unlikely the possibility that changes in the amount of enzyme protein were involved. Secondly, a direct effect on one of the enzymes of fatty acid synthesis must have been demonstrated to rule out the possibility that the effect of the hormone was not merely due to a change in substrate supply.

With respect to the first criterion, all of the effects quoted attained statistical significance within 1 h, some being detectable within 10 min. In several cases, indicated by asterisks, it was shown by immunological or enzymatic methods that the total amount of the rate-limiting enzyme, acetyl-CoA carboxylase, did not change.

With respect to the second criterion, a direct effect on one of the enzymes of fatty acid synthesis was shown in one of two ways. In some cases the hormone affected the incorporation of tritium from ^3H_2O into fatty acid without affecting incorporation into cholesterol. If fatty acids and cholesterol are assumed to be synthesized from the same cytoplasmic pool of acetyl-CoA, this makes it unlikely that the observed effect is the result of a change in the supply of substrate. In other cases a direct effect on acetyl-CoA carboxylase activity was observed in freshly-prepared extracts.

Even after these rather stringent criteria have been met, inspection of Table 1 reveals that a number of different hormones can cause rapid changes in the rate of fatty acid synthesis in intact cells. Measurements of the relative activities of acetyl-CoA carboxylase and fatty acid synthase in extracts suggest that the reaction catalysed by acetyl-CoA carboxylase would normally be rate-limiting [1]. In addition there is good evidence (see Sections 3 and 4) that the activity of acetyl-CoA carboxylase can be modulated in vitro by allosteric effectors and by reversible phosphorylation. It therefore seems likely that all of the effects listed in Table 1 will ultimately be explained by alteration of the activity of acetyl-CoA carboxylase.

It is worth pointing out that several of the hormones shown in Table 1 can also affect that rate of glycogen synthesis either in vivo or in intact cell systems. Glucagon [18], β-adrenergic [19], α-adrenergic agents [20], and vasopressin [21], all cause inhibition of glycogen synthesis, while insulin [22] causes a stimulation. Those hormones which inhibit glycogen synthesis cause a decrease in the activity of glycogen synthase measured in the absence of glucose-6-phosphate, while conversely those hormones which stimulate glycogen synthesis increase the activity of glycogen synthase in the absence of glucose-6-phosphate. The activity measured in the presence of glucose-6-phosphate is unchanged in all cases. The activity ratio ($-$ / $+$ glucose-6-phosphate) is a measure of the state of phosphorylation of glycogen synthase [23], and the implication is that all of these hormones exert their effects on glycogen synthesis by changing the state of phosphorylation of glycogen synthase. It is obviously an attractive hypothesis that the effects of these hormones on fatty acid synthesis could be mediated via changes in the state of phosphorylation of acetyl-CoA carboxylase. A logical extension of this hypothesis would be that the same protein kinases and phosphatases, regulated by the same effector molecules, might act on both glycogen synthase and acetyl-CoA carboxylase. This idea will be discussed further in Section 6.

In summary, there is now good evidence that fatty acid synthesis can be regulated in the short term by hormones via modulation of the activity of acetyl-CoA carboxylase. It remains to be established whether this regulation is achieved via changes in the concentration of allosteric effectors or by covalent modifications of the protein such as phosphorylation–dephosphorylation. Previous reviews have tended to stress the former type of regulation to the virtual exclusion of the latter. It is hoped that this article will redress the balance a little.

3. Acetyl–CoA carboxylase – structure and allosteric regulation

3.1. Structure

Acetyl-CoA carboxylase has been purified to apparent homogeneity from chicken liver [24], rat liver [25,26], bovine adipose tissue [27], rat mammary gland [28] and rabbit mammary gland [29,30]. The active form of the enzyme from each of these sources is a long, linear polymer with a molecular weight of about 10^7. In sedimentation velocity experiments a hypersharp boundary is obtained with a sedimentation coefficient, $s_{20,w}$ of around 50 S [25–28,30,31]. The polymers can be visualised in the electron microscope by negative staining and appear as long, narrow filaments which appear to have been formed by a helical aggregation of globular subunits [31]. Light-scattering [32] and viscosity [33] studies are also consistent with this highly asymmetric structure.

Acetyl-CoA carboxylase from *Escherichia coli* can be resolved into three separate proteins which have been purified to homogeneity [34]. One of these is the carboxyl carrier protein (CCP) which bears the biotinyl prosthetic group, and the other two are enzymically active proteins which catalyse the following partial reactions:

(1) Biotin carboxylase
$$ATP + CO_2 + biotin\text{-}CCP \rightarrow CO_2\text{-}biotin\text{-}CCP + ADP + P_i$$

(2) Carboxyl transferase
$$CO_2\text{-biotin-CCP} + \text{acetyl-CoA} \rightarrow \text{biotin-CCP} + \text{malonyl-CoA}$$

Resolution of animal acetyl-CoA carboxylase into these three components has not been achieved, although the partial reactions can be demonstrated by isotopic exchange reactions [35]. This failure to separate the activities was originally taken to mean that animal acetyl-CoA carboxylases existed as tightly bound multienzyme complexes. Early results using polyacrylamide gel electrophoresis in the presence of sodium dodecyl sulphate suggested that chicken liver [36] and rat liver [26] acetyl-CoA carboxylase contained up to four different subunits with molecular weights in the range of 120 000, and it was suggested that these subunits represented the different enzymatic components of the multienzyme complex. However, it is now clear that the apparent 'non-identical' subunits of animal acetyl-CoA carboxylases are artefacts due to proteolysis of the enzyme during its preparation. This was first suggested by Tanabe et al. [37], who found that the enzyme purified from rat liver sometimes gave a single species of molecular weight 230 000 when analysed by polyacrylamide gel electrophoresis in the presence of sodium dodecyl sulphate. On other occasions they obtained two species with molecular weights of 118 000 and 125 000, and on some occasions a mixture of all three species. Both the 230 000 and 125 000 molecular weight species contained covalently bound biotin, and the 230 000 species could be converted into a mixture of the 125 000 and 118 000 dalton species by treatment with chymotrypsin or trypsin. The inference was that the intact acetyl-CoA carboxylase subunit (molecular weight 230 000) was subject to variable degrees of proteolysis during the preparation of the enzyme. Since this original report, acetyl-CoA carboxylase has been isolated by direct immunoprecipitation of extracts of rat mammary gland [38] chicken liver [38], and rat adipose tissue [39] and found to consist of a single 230 000 molecular weight species. Acetyl-CoA carboxylases purified recently from rat [28] and rabbit [30] mammary glands also appear to consist of single subunits of molecular weight 250 000. Taken together, these results show that animal acetyl-CoA carboxylase consists of a single type of subunit of molecular weight 230 000–250 000, and is, therefore, an example of a multifunctional protein [40] with two active sites on one polypeptide chain. The biotin contents (1 mol/240 000 g protein) obtained for the rat liver [26,37] and rat mammary gland [28] enzymes are consistent with this model. Whether the various values estimated for the molecular weight of the subunit (215 000 → 250 000 [26,28,30,37–39] reflect experimental error or genuine differences is not yet clear. The only two studies in which marker proteins of molecular weight comparable to or larger than acetyl-CoA carboxylase were used, gave values of 250 000 for the carboxylase subunit [28,30] and this will be assumed to be the correct subunit molecular weight for the remainder of this article.

The polymeric form of acetyl-CoA carboxylase ($s_{20,w}$ = 50 S) can be converted into an inactive 'protomeric' form ($s_{20,w}$ = 13 S) by treatment with high ionic strength at alkaline pH, or by incubation with substrates in the absence of the allosteric activator, citrate [41]. Sedimentation equilibrium experiments suggested a molecular weight of 410 000 for the protomeric form of chicken liver acetyl-CoA carboxylase [41] and 560 000 for the equivalent form of the bovine adipose tissue enzyme [27]. Tanabe et al. [37] reported that the 'protomeric' form of rat liver acetyl-CoA carboxylase gave curved plots in sedimentation equilibrium experiments which could be explained by an association–disassociation equilibrium between the 250 000 dalton subunit and its dimer. The basic unit of acetyl-CoA carboxylase which aggregates into the polymeric form may therefore be a dimer.

Studies on the chicken liver enzyme by Gregolin et al. [42] suggested that the dimer may also be the minimal binding unit for acetyl-CoA carboxylase. They reported that there were 0.6 binding sites for bicarbonate and acetyl-CoA, 0.7 binding sites for citrate, and 0.6 molecules of biotin per 250 000 daltons. However, since three other laboratories [26,28,37] have reported that the mammalian enzyme contains 1.0 molecules of biotin per 250 000 dalton subunit, the number of binding sites needs to be re-examined. Apart from the data of Gregolin et al. [42], all available evidence points to the conclusion that there is only one type of subunit.

In summary, acetyl-CoA carboxylase can exist in vitro as an active, polymeric form or an inactive protomeric form. The latter form is probably a dimer of identical subunits of molecular weight 250 000. Whether the polymer ⇌ protomer equilibrium operates in vivo is an open question, although Meredith and Lane [43] have recently obtained some indirect evidence that it does. If chicken liver cells in monolayer culture are exposed to digitonin, a treatment which perforates the plasma membranes without disrupting the cell, acetyl-CoA carboxylase leaks out much more slowly than other soluble enzymes such as lactate dehydrogenase. However pretreatment of the cells with dibutyryl cyclic AMP, which is considered to promote dissociation to the protomeric form by lowering the cytoplasmic citrate concentration [5], causes acetyl-CoA carboxylase to leak out rapidly after digitonin treatment. If malonyl-CoA or citrate are included with the digitonin, the rate of leakage is affected in the manner predicted from the known effects of these molecules on the polymer ⇌ protomer equilibrium. In addition the slowly leaking form is not inhibited by the biotin-binding protein avidin. This is consistent with the idea that it represents the polymeric form of the enzyme.

3.2. Allosteric control

Animal acetyl-CoA carboxylases have an absolute requirement for citrate or isocitrate, and are completely inactive in the absence of these activators. The effect of citrate on fatty acid synthesis was first noted by Brady and Gurin [44] in 1952, and 10 years later it was shown by Vagelos et al. [45] that the effect could be explained by the activation of acetyl-CoA carboxylase. The activation was accompanied by an increase in the sedimentation rate of the enzyme in sucrose gradients [46], and it was subsequently shown, using the purified chicken liver enzyme [41], that citrate activation was associated with the conversion of the protomer to the polymeric form. It is still not clear whether polymerisation is a prerequisite or merely a consequence of the activation, although there is a strict temporal correlation between inactivation and depolymerisation if the latter is assessed by viscosity measurements [47]. Citrate increases the V_{max} rather than decreasing the K_m values for substrates [24,48], and it activates both partial reactions [35]. The reader is referred to the review by Lane et al. [49] for a full discussion of the mechanism of citrate activation.

Acetyl-CoA is formed in mitochondria by β-oxidation of fatty acids or by the pyruvate dehydrogenase reaction. The carbon atoms of acetyl-CoA can only be transported into the cytoplasm in the form of citrate, which is therefore a precursor for fatty acid synthesis. The concentrations of citrate estimated to exist in the cytoplasm ($0.2 \rightarrow 2$ mM) [50] are of the same order as the concentration required for half-maximal activation of acetyl-CoA carboxylase in vitro. It has, therefore, been proposed that citrate is a feed-forward activator of fatty acid synthesis which activates acetyl-CoA carboxylase when the levels of tricarboxylic acid cycle intermediates are high.

Unfortunately attempts to correlate levels of citrate with rates of fatty acid synthesis in vivo have met with varying success. In one study reported by Brownsey et al. [51], treatment of rat fat pads with fluoroacetate (which is metabolised into fluorocitrate, an inhibitor of aconitate hydratase) increased the tissue concentration of citrate 10-fold, but did not affect the rate of fatty acid synthesis or the activity of acetyl-CoA carboxylase. A drawback of this and numerous other studies was that only the average tissue concentration of citrate was estimated, despite the fact that citrate is distributed between the cytoplasm and the mitochondrial matrix. Using new methods for rapid cell fractionation, Watkins et al. [5] have shown that concentrations of glucagon which inhibit fatty acid synthesis in cultured chick liver cells decrease the cytoplasmic citrate concentration by an order of magnitude. This study represents the most convincing correlation to date between the rate of fatty acid synthesis and the concentration of citrate in the cell. It does not, however, rule out the possibility that glucagon acts on acetyl-CoA carboxylase by additional mechanisms, such as phosphorylation.

In 1963, Bortz and Lynen [52] reported that palmityl-CoA and other long-chain acyl-CoA derivatives inhibited acetyl-CoA carboxylase activity. Since fatty acyl-CoA can be regarded as an end-product of fatty acid synthesis (after re-esterification of the free fatty acid product by fatty acyl-CoA ligase), it was suggested that this represented a feedback control of fatty acid synthesis. This could be an important mechanism for the regulation of de novo fatty acid synthesis by dietary fatty acids. It was found subsequently that fatty acyl-CoA was competitive with respect to citrate, and favoured the conversion of the enzyme to the inactive, protomeric form [48].

The significance of palmityl-CoA as a feedback control mechanism has been challenged on the grounds that long-chain fatty acyl-CoA is a detergent which inhibits many enzymes other than those involved in fatty acid synthesis [53]. The concentrations of palmityl-CoA required to produce inhibition of acetyl-CoA carboxylase are at or near the concentration required to form micelles [54].

In answer to these criticisms, Goodridge [55] showed that palmityl-CoA inhibited acetyl-CoA carboxylase in chicken liver extracts even in the presence of high concentrations of serum albumin, conditions where other detergent-inhibited enzymes were not affected. The palmityl-CoA effect has also been placed on firmer ground by some recent work of Ogiwara et al. [56], who showed that purified rat liver acetyl-CoA carboxylase bound 1 molar of palmityl-CoA per subunit, with a K_i of 5.5 nM. If the molecular ratio palmityl-CoA/enzyme was kept low (<2:1), a reversible depolymerisation and inactivation of the enzyme was observed, which was associated with the binding of 1 molecule of palmityl-CoA. If the molar ratio palmityl-CoA/enzyme was > 5:1, another type of inhibition was observed which was associated with aggregation of the enzyme to forms that sedimented more rapidly than 50 S. This latter effect could well be the non-specific 'detergent' effect observed by several authors.

Unfortunately, the estimation of cytoplasmic long-chain fatty acyl-CoA is even more difficult than the estimation of cytoplasmic citrate. Fatty acyl-CoA is not only distributed between cytoplasm and mitochondria, but most, if not all of it is also protein bound. It is therefore not surprising that attempts to correlate rates of fatty acid synthesis with levels of long-chain fatty acyl-CoA have met with limited success. Indeed it is difficult to see how it could be demonstrated conclusively that palmityl-CoA inhibition of acetyl-CoA carboxylase operates in vivo.

In summary, while the effects of both citrate and palmityl-CoA on acetyl-CoA carboxylase activity are now well documented in vitro, evidence that these controls operate in vivo has been difficult to obtain.

4. Acetyl-CoA carboxylase – phosphorylation and dephosphorylation

4.1. Studies in partially purified systems

The first hint that acetyl-CoA carboxylase might be regulated by phosphorylation came from the work of Inoue and Lowenstein, who reported that the purified rat liver enzyme contained 2.4 molecules of phosphate per subunit of molecular weight 250 000 [26]. However the first direct evidence for this hypothesis was provided by Carlson and Kim [57]. These workers prepared a high-speed supernatant from a rat liver extract, and then partially purified acetyl-CoA carboxylase by collecting the protein which precipitated at 40% saturated ammonium sulphate. If this fraction was incubated with magnesium ions, a time-dependent activation of acetyl-CoA carboxylase was observed. This activation was blocked by 50 mM sodium fluoride, which is known to be an inhibitor of protein phosphatases. If the fraction was incubated with MgATP plus another protein fraction which precipitated between 65% and 75% saturated ammonium sulphate (fraction K), a time-dependent inactivation of acetyl-CoA carboxylase was observed. This inactivation was not observed in the absence of either fraction K or MgATP [58]. Radioactivity was incorporated from $[\gamma\text{-}^{32}P]ATP$, but not $[U\text{-}^{14}C]ATP$, into protein which could be precipitated using an antibody raised against purified rat liver acetyl-CoA carboxylase.

These results showed that rat liver acetyl-CoA carboxylase could be converted from a high activity form (carboxylase *a*) to a low activity form (carboxylase *b*), by incubation in vitro with MgATP, and that this inactivation was accompanied by phosphorylation of the protein. In a subsequent paper [59] Carlson and Kim compared the kinetic properties of these two forms of the enzyme in more detail. The dependence on citrate is illustrated in Figure 1. The *a* form was five times more active than the *b* form, even at saturating citrate concentrations. However, the dependence on citrate concentration of the *b* form was sigmoidal and the concentration of citrate required for half-maximal activation was 2.4 mM as opposed to 0.2 mM for

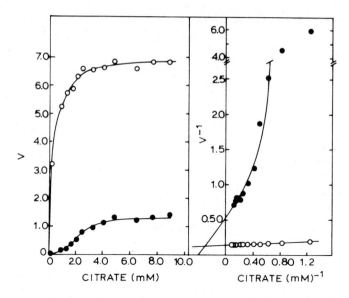

Figure 1. Effect of the concentration of citrate in the assay, on the activity of acetyl-CoA carboxylase partially purified from rat liver. ○——○ acetyl-CoA carboxylase *a*; ●——● acetyl-CoA carboxylase *b* (i.e., after inactivation by treatment with MgATP). (Figure reproduced with permission from Carlson and Kim [59].)

the *a* form. Thus at physiological citrate concentrations (0.2 → 2 mM) [50] the *a* form would be at least an order of magnitude more active than the *b* form. The two forms of acetyl-CoA carboxylase had similar K_m values for the substrates acetyl-CoA and ATP but differed markedly in V_{max}. The *b* form was also much more sensitive to inhibition by palmityl-CoA although it is not clear whether or not this was the non-specific detergent effect described in Section 3.2.

The results of Carlson and Kim indicated that acetyl-CoA carboxylase could be interconverted by a phosphorylation–dephosphorylation mechanism between two forms which differed markedly in their kinetic properties. It was proposed that this mechanism could be involved in the short-term regulation of fatty acid synthesis by insulin and adrenaline.

The conclusions of Carlson and Kim were initially challenged on several counts. In particular the inactivation they observed was attributed to mechanisms other than phosphorylation. There are, of course, many possible reasons for the inactivation of an enzyme in a rather impure system, e.g., proteolysis. It has also been pointed out [60,61] that there is a mechanism by which even homogeneous acetyl-CoA carboxylase can become inactivated on incubation with MgATP. It is well known that the carboxylated intermediate form of acetyl-CoA carboxylase (Section 3.1) tends to dissociate into the inactive, protomeric form in the absence of citrate [41]. Hence the inactivation observed by Carlson and Kim could theoretically be explained by this effect, a slow carboxylation of the enzyme being brought about in the presence of MgATP by traces of dissolved carbon dioxide in the medium. Indeed Kim and coworkers [62] subsequently reported that conversion of the *a* form to the *b* form was associated with depolymerisation of the enzyme as judged by sucrose density gradient analysis, which might appear to lend support to this view. However, Kim and coworkers [62] could observe inactivation with MgATP even in the presence of low concentrations of citrate and in buffers which were free of carbon dioxide. In any case, carboxylation of the enzyme cannot explain Carlson and Kim's early reports [57,58] that the inactivation of acetyl-CoA carboxylase by MgATP was completely dependent on the addition of protein fraction K.

A time-dependent inactivation of acetyl-CoA carboxylase by MgATP has also been observed by other workers in partially purified preparations from bovine mammary gland [63] and rat liver [64]. In the latter case the authors suggested that the inactivation was not due to phosphorylation, since they could separate the bulk of the phosphorylated protein from acetyl-CoA carboxylase activity by chromatography on DEAE-cellulose. However since the phosphorylation reaction was carried out with a high-speed supernatant of a liver homogenate, acetyl-CoA carboxylase would only represent a minute proportion of the total phosphoprotein, and this criticism is not valid.

The conclusions of Carlson and Kim have also been challenged by Lane et al. [61] on the grounds that the phosphorylated protein precipitated by the antibody to acetyl-CoA carboxylase may not have been acetyl-CoA carboxylase. This possibility can be seriously entertained since Walker et al. [65] have found that antibody raised against mammary gland acetyl-CoA carboxylase requires extensive purification by immunoabsorption procedures to render it monospecific. However, Lee and Kim [65] subsequently examined ^{32}P-labelled immunoprecipitates by polyacrylamide gel electrophoresis in the presence of sodium dodecyl sulphate and found that the ^{32}P-radioactivity was mainly associated with a polypeptide of molecular weight 215 000, with smaller amounts in a polypeptide of molecular weight 125 000, which represents the larger proteolytic fragment of the enzyme. There thus seems little doubt that acetyl-CoA carboxylase is phosphorylated in these experiments.

The ATP-dependent inactivation of acetyl-CoA carboxylase was originally reported to be unaffected by cyclic AMP in the range of $10^{-7}M$ to $10^{-4}M$

[57]. Carlson and Kim therefore suggested that fraction K represented a cyclic AMP-independent protein kinase. However, Kim and co-workers [62] subsequently reported that if the enzyme preparation was preincubated with cyclic AMP before addition of ATP, it did stimulate both phosphorylation and inactivation. Since the concentration of cyclic AMP which was required for maximal stimulation (10^{-4} M) was at least twenty times higher than that required for the maximal activation of cyclic AMP-dependent protein kinase [67], the significance of this observation is not clear.

It seems reasonable to conclude from the data of Kim and coworkers that MgATP does cause inactivation of acetyl-CoA carboxylase and that this is associated with the phosphorylation of the enzyme. However in order to conclude that phosphorylation per se causes inactivation, it is necessary to demonstrate that complete inactivation correlates with the incorporation of stoichiometric amounts of phosphate into the protein. Reliable estimates for the stoichiometry of phosphorylation can only be obtained by using the purified enzyme. This is the approach that has been adopted in this laboratory and the results are discussed in the following section.

4.2. Studies with the purified enzyme

In this laboratory we have chosen to work with acetyl-CoA carboxylase from lactating rabbit mammary gland. Purification of the enzyme from this tissue has two advantages over the traditional sources, i.e., liver and adipose tissue. Firstly the enzyme is present in much larger amounts: only 200–400-fold purification is required to obtain homogeneous acetyl-CoA carboxylase from lactating rabbit mammary gland [68], as against 1200–2000-fold purification for the enzyme from liver [24–26] or adipose tissue [27]. Secondly, far fewer problems of proteolysis are encountered with the lactating mammary gland, as discussed by Hardie and Cohen [30].

We have purified acetyl-CoA carboxylase from lactating rabbit mammary gland by a novel procedure that involves only precipitation with ammonium sulphate and polyethylene glycol [68]. The results of a typical purification procedure are shown in Table 2. The enzyme activity is purified 300-fold in 40% yield, with up to 10 mg of homogeneous acetyl-CoA carboxylase being prepared from each rabbit. The preparation is homogeneous by a variety of electrophoretic and ultracentrifugal criteria [30]. In sedimentation velocity experiments a single, hypersharp sedimenting boundary [50 S] is obtained, which corresponds to the polymeric form of the enzyme. On polyacrylamide gel electrophoresis in the presence of sodium dodecyl sulphate, a single polypeptide with a molecular weight of 250 000 is observed. Traces of lower

Purification of acetyl-CoA carboxylase from lactating rabbit mammary gland. 300 g tissue (2 rabbits) was used in this preparation

Step	Activity	Protein	Specific activity	Purifi- cation	Yield
1. 90 000 × g supernatant	168	12,400	0.014	1	100
2. 35% ammonium sulphate precipitate	176	4,800	0.037	2.6	105
3. 1st polyethylene glycol precipitate	132	90	1.5	105	79
4. 2nd polyethylene glycol precipitate	85	20	4.2	300	51

molecular weight polypeptides (240 000 and 230 000) in this preparation are now known to be due to slight proteolysis during the preparation of the enzyme, and can be eliminated by the inclusion of proteinase inhibitors in the purification media (P. Guy and D.G. Hardie, unpublished observations). The purified enzyme contains a substantial amount of covalently bound alkali-labile phosphate. We originally reported an average value of 3.2. molecules of phosphate per subunit based on three preparations of the enzyme [30]. More recently however, a much larger number of preparations have averaged 5 molecules of phosphate per subunit [69] and a value of 6.4 molecules of phosphate per subunit has been reported for the rat mammary gland enzyme [28].

Despite the fact that the preparation is essentially homogeneous, it is still contaminated with traces of protein kinases which can phosphorylate the 250 000 molecular weight subunit of acetyl-CoA carboxylase [68]. If the purified enzyme is incubated with magnesium ions and $[\gamma\text{-}^{32}P]ATP$ in the absence of cyclic AMP, there is a phosphorylation of the enzyme which reaches a value of up to 0.7 molecules of phosphate per subunit in 60 min (Figure 2). If the incubation is carried out in the presence of cyclic AMP, a more rapid phosphorylation is observed which reaches a value of 1.5 molecules per subunit in 60 min (Figure 2). The cyclic AMP-stimulated phosphorylation is clearly due to traces of endogenous cyclic AMP-dependent protein kinase. However the basal phosphorylation observed in the absence of cyclic AMP could be due either to a cyclic AMP-independent protein kinase, or to the free catalytic subunit of cyclic AMP-dependent protein kinase, since the latter is active in the absence of cyclic AMP [70]. These two possibilities

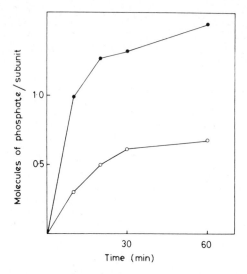

Figure 2. Phosphorylation of acetyl-CoA carboxylase by endogenous protein kinases. Acetyl-CoA carboxylase was incubated at 1.3 mg/ml with 0.021 Tris/HCl buffer (pH 7.2), 0.4 mM EDTA, 0.1 mM EGTA, 0.4 mM dithiothreitol, 3 mM Mg acetate, 1 mM [γ-^{32}P]ATP, in the presence ($\bullet$) or absence ($\bigcirc$) of 10 μM cyclic AMP.

were tested using the heat-stable inhibitor protein of cyclic AMP-dependent protein kinase (see Chapter 1). Addition of an excess of this protein completely abolished the cyclic AMP-stimulated phosphorylation, but did not affect the basal phosphorylation [68]. Acetyl-CoA carboxylase is therefore phosphorylated by at least two protein kinases; i.e., cyclic AMP-dependent protein kinase and a cyclic AMP-independent protein kinase which we have provisionally termed acetyl-CoA carboxylase kinase-2 [68]. These two protein kinases differ in their sensitivity to the inhibitor protein and also in their K_m for ATP, acetyl-CoA carboxylase kinase-2 having a value for K_m around ten times higher than that of cyclic AMP-dependent protein kinase (D.G. Hardie, unpublished observations).

The finding that an apparently homogeneous enzyme is contaminated with protein kinase is perhaps initially surprising. However, experience with other enzymes that are known to be controlled by phosphorylation, such as phosphorylase kinase [71] and glycogen synthase [72], has shown that it is extremely difficult to remove all traces of protein kinases from the purified enzymes. It is worth pointing out that cyclic AMP-dependent protein kinase was first discovered as an endogenous contaminant in highly purified preparations of phosphorylase kinase [73]. The contamination of purified acetyl-CoA carboxylase by cyclic AMP-dependent protein kinase can be estimated at no more than 0.01% by weight, which is well below the limits of sensitivity of any physicochemical criterion of purity. This emphasizes the importance of characterizing these contaminating activities even in an apparently homogeneous protein.

Acetyl-CoA carboxylase can also be phosphorylated rapidly by adding the catalytic subunit of cyclic AMP-dependent protein kinase purified from skeletal muscle [68]. In Table 3 the rate of phosphorylation of acetyl-CoA carboxylase by the catalytic subunit is compared with the rates of

TABLE 3

Phosphorylation of enzymes involved in carbohydrate and lipid metabolism by cAMP-dependent protein kinase. Results are expressed relative to the rate of phosphorylation of the β-subunit of phosphorylase kinase

Protein	Initial rate of phosphory-lation (%)
Phosphorylase kinase (β-subunit)	100
Glycogen synthase (site 2)	80
Acetyl-CoA carboxylase (mammary gland)	35
Pyruvate kinase (L type) (liver)	35
Protein phosphatase inhibitor-1	30
Phosphorylase kinase (α-subunit)	20
Phosphorylase b	0.02
Glycogen debranding enzyme	0.01
Phosphoglucomutase	0.03
Hexokinase (type 1)	0
Glucose-6-Phosphate dehydrogenase	0.01
Phosphohexose isomerase	0
Phosphofructokinase	0.1
Fructose-1,6-bisphosphatase (liver)	0.2
Aldolase	0.01
Triose phosphate isomerase	0.01
Glyceraldehyde-3-Phosphate dehydrogenase	0.01
Phosphoglycerate kinase	0
Phosphoglyceromutase	0
Enolase	0.02
Pyruvate kinase (M type)	0
Lactate dehydrogenase	0.01
Creatine kinase	0.01
Fatty acid synthase (mammary gland)	0.01

Full details of the incubations are described by Cohen [74]. Owing to the low solubility of purified acetyl-CoA carboxylase, its phosphorylation was measured relative to the β-subunit of phosphorylase kinase at a protein concentration of 1.5 μM rather than 6 μM as used for all other proteins. Enzymes were from rabbit skeletal muscle unless stated otherwise.

phosphorylation of various other proteins investigated by Cohen and coworkers [30,74]. It can be seen that acetyl-CoA carboxylase is phosphorylated at rates comparable to other well-characterized substrates of cyclic AMP-dependent protein kinase such as phosphorylase kinase, glycogen synthase and liver pyruvate kinase. In addition, while cyclic AMP-dependent protein kinase can phosphorylate several different proteins, it must not be regarded as a non-specific protein kinase. Acetyl-CoA carboxylase is phosphorylated at least 1000 times faster than any of the glycolytic enzymes from muscle, or fatty acid synthase from lactating rabbit mammary gland. Acetyl-CoA carboxylase thus belongs to a small group of key regulatory enzymes that are good substrates for cyclic AMP-dependent protein kinase.

We have recently carried out a preliminary analysis of the sites phosphorylated on acetyl-CoA carboxylase by cyclic AMP-dependent protein kinase and acetyl-CoA carboxylase kinase-2 (D.G. Hardie and P. Cohen, unpublished results). Acetyl-CoA carboxylase was phosphorylated using the endogenous protein kinases under conditions where only one kinase would be active (presence of cyclic AMP and low ATP concentration for cyclic AMP·dependent protein kinase; absence of cyclic AMP, presence of the inhibitor protein, and high ATP concentration for acetyl-CoA carboxylase kinase-2). The phosphorylated protein was digested with trypsin and the resultant peptides analysed by gel filtration or two-dimensional peptide mapping. The results of autoradiography of the peptide maps are shown in Figure 3. Acetyl-CoA

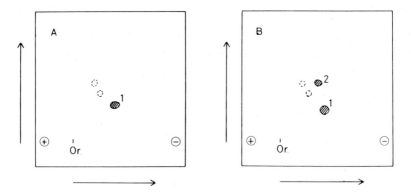

Figure 3. Autoradiograms of two-dimensional maps of tryptic peptides of acetyl-CoA carboxylase which had been phosphorylated by acetyl-CoA carboxylase kinase-2 (A) or cyclic AMP-dependent protein kinase (B). Samples were applied to cellulose thin layer plates at the origin (Or.) and electrophoresed in the horizontal dimension in pyridine/acetic acid/H_2O (1 : 10 : 189) (pH 3.5). Chromatography was carried out in the vertical dimension using butanol/pyridine/acetic acid/H_2O (90 : 60 : 18 : 72) as eluant. Minor spots (accounting for <10% of the total radioactivity) are indicated by dotted lines.

carboxylase kinase-2 phosphorylates a single major site which we have shown to be a serine residue (site 1). Cyclic AMP-dependent protein kinase phosphorylates serine residues which occur in two tryptic peptides. One of these peptides appears to be identical to the peptide containing site 1 since it runs identically on electrophoresis, chromatography and gel filtration. Since this peptide is probably only a few residues long (K_{AV} on Sephadex G-50 = 0.79), it seems likely that both acetyl-CoA carboxylase kinase-2 and cyclic AMP-dependent protein kinase phosphorylate the same serine residue, i.e., site 1. The other phosphorylation site is exclusive to cyclic AMP-dependent protein kinase and will be referred to as site 2. Acetyl-CoA carboxylase kinase-2 and cyclic AMP-dependent protein kinase thus have different phosphorylation site specificities. This finding will enable investigation as to which of these protein kinases is operating under different conditions in vivo.

Before the phosphorylation of a protein can be considered to be physiologically relevant, it is necessary to show that the phosphorylation correlates with a change in the function of the protein. Fulfilment of this criterion has become especially important since the demonstration by Bylund and Krebs [75] that proteins can become substrates for cyclic AMP-dependent protein kinase when they are denatured, even when the native protein is not a substrate.

We have recently prepared a highly phosphorylated form of acetyl-CoA carboxylase which is activated on dephosphorylation [69]. If our normal purification procedure was modified by including 50 mM sodium fluoride in all the buffers, the purified enzyme contained 6 mol of phosphate per subunit as against 4–5 mol per subunit for the normal preparation. The higher phosphate content correlated with a lower specific activity of the purified enzyme, i.e., 1.2 → 1.5 Units/mg as against 3 → 4 Units/mg for the normal preparation. The low-specific activity form could be converted back to the high-specific activity form by treatment with highly-purified protein phosphatase-1 from skeletal muscle. Protein phosphatase-1 is the enzyme which dephosphorylates sites on phosphorylase, phosphorylase kinase and glycogen synthase, and thus converts these enzymes into forms which favour glycogen synthesis rather than glycogen breakdown (Chapter 1). If acetyl-CoA carboxylase was assayed at 1 mM citrate (Figure 4 B), protein phosphatase-1 caused a 4-fold activation of enzyme prepared in the presence of fluoride. At the same time the alkali-labile phosphate content was reduced from 5.9 to 4.4 molecules per subunit. Under the same conditions acetyl-CoA carboxylase prepared by the normal procedure was activated only 1.4-fold (Figure 4 A) and the phosphate was reduced from 4.9 to 4.3 molecules per subunit.

50

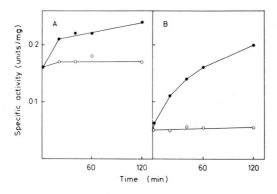

Figure 4. Activation of acetyl-CoA carboxylase by incubation with protein phosphatase-1. Acetyl-CoA carboxylase prepared by the normal procedure (A) or in the presence of 50 mM NaF (B) was incubated either with (●) or without (○) protein phosphatase-1 (2 units/ml) in 0.04I Tris/HCl buffer (pH 7.5), 0.8 mM EDTA, 0.8 mM dithiotreitol, 2 mM $MnCl_2$. At intervals aliquots were withdrawn for the assay of acetyl-CoA carboxylase activity at 1 mM citrate, pH 7.2. Reproduced with permission from [69].

   These results can be explained by proposing that protein phosphatase-1 dephosphorylates a site on acetyl-CoA carboxylase that affects the activity of the enzyme. This site is almost completely dephosphorylated during the preparation of acetyl-CoA carboxylase, unless fluoride is present in the buffer. This dephosphorylation occurs during the dialysis of the ammonium sulphate fraction, since a large activation is observed at this step [69]. Two additional findings provide conclusive evidence that the activation produced by protein phosphatase-1 is due to dephosphorylation per se and not to some other property of the phosphatase preparation. Firstly the activation was completely dependent on the presence of manganese ions, and the dephosphorylation of other phospho-protein substrates of this preparation of protein phosphatase-1 is also stimulated by manganese ions [74]. Secondly, the activation is completely blocked by an excess of inhibitor-2. Inhibitor-2 is a heat-stable protein from skeletal muscle which inhibits protein phosphatase-1 but not other protein phosphatases that have so far been tested [74].
   There is therefore very good evidence that dephosphorylation, and by inference phosphorylation, modulates the activity of purified acetyl-CoA carboxylase in vitro. The effect is about 2-fold at saturating citrate concentrations, but is at least 4-fold at low citrate concentrations [69]. At the time of writing, experiments are in progress to determine the effects of phosphory-

lation on the kinetic properties of the enzyme. Preliminary results indicate that phosphorylation by either of the endogenous protein kinases (cyclic AMP-dependent or acetyl-CoA carboxylase kinase-2) does cause inactivation of acetyl-CoA carboxylase. However the amount of phosphorylation and the inactivation produced by either of these two enzymes is similar in control preparations and preparations prepared in the presence of fluoride, and is unaffected by prior treatment with protein phosphatase-1. The activation produced by protein phosphatase-1 therefore appears to be at a site distinct from both sites 1 and 2, and the phosphorylation of this site is presumably catalysed by a third protein kinase, as yet uncharacterized.

4.3. Studies with intact cell systems

The phosphorylation of acetyl-CoA carboxylase in intact cells was first demonstrated by Brownsey et al. [76]. If rat epididymal fat pads were incubated with ^{32}P-labelled inorganic phosphate, ^{32}P-radioactivity was incorporated into at least eight polypeptides, one of which had an apparent molecular weight of 230 000. The identity of the 230 000 molecular weight species with acetyl-CoA carboxylase was established by two independent methods. Firstly this polypeptide was precipitated quantitatively using antiserum raised against acetyl-CoA carboxylase, but not by control serum. Secondly, the ^{32}P-labelled polypeptide could be separated from other ^{32}P-labelled polypeptides by chromatography on avidin–Sepharose. Brownsey et al. [76] also reported that insulin did not significantly affect the phosphorylation of acetyl-CoA carboxylase. However, as it is now clear that the enzyme is phosphorylated at multiple sites (Section 4.2), measurement of the overall level of phosphorylation may not be sufficiently sensitive to detect a small change in the state of phosphorylation of one site.

Since this initial report, similar techniques have been used to demonstrate the phosphorylation of acetyl-CoA carboxylase in intact liver cells. Avruch et al. [77] showed that glucagon enhanced the phosphorylation of several polypeptides in rat liver cells, including one polypeptide of molecular weight 220 000. Witters et al. [78] identified this polypeptide as acetyl-CoA carboxylase by immunoprecipitation, and showed that glucagon enhanced its phosphorylation by 50%. Although they did not show that this increased radioactivity was also precipitated by the antibody, glucagon treatment did cause a 40% decrease in the activity of acetyl-CoA carboxylase measured in an extract of the cells.

Acetyl-CoA carboxylase is also phosphorylated in chick liver cells maintained in monolayer culture [79]. Radioactivity is incorporated from ^{32}P-labelled phosphate in the medium into the 240 000 molecular weight subunit of acetyl-CoA carboxylase. A stoichiometry at steady state of 9 to 10 molecules of phosphate per subunit was calculated. However this estimate assumes that the specific activity of acetyl-CoA carboxylase in the extract is identical to that of the purified enzyme, and that the specific radioactivity of cytoplasmic ATP was identical with that of the phosphate in the medium, and could be considerably in error. The authors also reported that treatment of the cells with dibutyryl cyclic AMP, which markedly inhibits fatty acid synthesis in these cells [5], did not change either the activity or the ^{32}P content of acetyl-CoA carboxylase. This is in contrast to the results of Witters et al. [78] with glucagon and rat hepatocytes, although in fact the two studies are not directly comparable. Witters et al. equilibrated rat hepatocytes with [^{32}P]phosphate and then incubated with glucagon for 5–15 min. Pekala et al. [79] added dibutyryl cyclic AMP at the same time as [^{32}P]phosphate and continued the experiment for 10 h to ensure that the labelling of acetyl-CoA carboxylase reached a steady state. This length of incubation would seem rather inappropriate for a study of short-term regulation by hormones. In fact, the data of Pekala et al. [79] do indicate a stimulation by dibutyryl cyclic AMP of the ^{32}P content of acetyl-CoA carboxylase at the early time points (1–3 h). While this stimulation was small, it represents $0.5 \rightarrow 1.0$ molecules of phosphate per subunit if their method for calculating the phosphorylation stoichiometry is valid.

The possibility that hormones which raise cyclic AMP bring about increased phosphorylation of acetyl-CoA carboxylase has recently been strengthened by reports from two laboratories on the effects of adrenaline on adipose tissue. Lee and Kim [80] have shown that adrenaline treatment doubles the phosphorylation of acetyl-CoA carboxylase within 30 min either in vivo or with isolated fat pads. This was accompanied by an inactivation of the enzyme which could be reversed by incubating the homogenate with magnesium ions, but was not reversed by incubation with citrate. This latter observation shows that the inactivation is not produced by high levels of palmityl-CoA formed by adrenaline-stimulated lipolysis, since palmityl-CoA inhibition is reversed by citrate [48]. The adrenaline-stimulated phosphorylation and inactivation was inhibited by the β-antagonist propranolol, while the basal phosphorylation was not affected. Lee and Kim took this as evidence that the basal and adrenaline-stimulated phosphorylation reactions were occurring at different sites on the enzyme. In support of this hypothesis they stated that the basal phosphorylation did not cause inactivation of acetyl-CoA carboxylase. However, since the authors had no way of preventing this basal phosphorylation, it is difficult to see how this statement can be justified.

While the results from our laboratory with the purified rabbit mammary gland enzyme would favour the idea that acetyl-CoA carboxylase can be phosphorylated at multiple sites (Section 4.2), further work is required to prove that this occurs in vivo.

Similar results have been reported by Brownsey et al. [81], who found that adrenaline stimulated the phosphorylation of acetyl-CoA carboxylase in isolated fat pads or cells by 50%. This was accompanied by a 50% decrease in the activity of the enzyme measured at saturating citrate concentrations, an effect that was not overcome by incubation for 30 min with 20 mM citrate. This is in contrast to the effects of insulin on acetyl-CoA carboxylase activity in isolated fat pads, which are reversed by preincubation with high concentrations of citrate [60]. Brownsey et al. [81] also reported that the incubation of a homogenate from adrenaline-treated cells with magnesium ions, or magnesium plus calcium ions, resulted in the loss of ^{32}P-radioactivity from acetyl-CoA carboxylase. This could be due either to dephosphorylation or to proteolysis. However, since it correlated with an activation of acetyl-CoA carboxylase back to the control value, dephosphorylation is the most likely explanation. The authors also compared the dependence of acetyl-CoA carboxylase activity on citrate concentration in extracts from control and adrenaline-treated cells. Adrenaline decreased the activity of acetyl-CoA carboxylase at all concentrations of citrate tested. There were also small changes in the concentration of citrate required for half-maximal activation. However it is difficult to assess the significance of this latter observation because there was a substantial amount of activity even in the absence of added citrate, while the purified enzyme is completely dependent on citrate (Section 3.2).

In summary, there is now convincing evidence that hormones which can raise cyclic AMP levels, i.e., glucagon in the liver and adrenaline in adipose tissue, can cause parallel inactivation and phosphorylation of acetyl-CoA carboxylase. A logical hypothesis is that both phosphorylation and inactivation are brought about by cyclic AMP-dependent protein kinase, since this enzyme is known to be capable of very rapid phosphorylation and inactivation of purified rabbit mammary gland acetyl-CoA carboxylase (Section 4.2).

As yet no hormones other than glucagon or adrenaline have been shown to alter the phosphorylation state of acetyl-CoA carboxylase. It has been reported that insulin does not change the state of phosphorylation of the enzyme in isolated fat tissue [75]. However in view of the accumulating evidence that acetyl-CoA carboxylase is phosphorylated at multiple sites, it will be necessary to examine the phosphorylation of individual sites before the lack of effect of insulin can be accepted.

4.4. Acetyl-CoA carboxylase phosphatase

There is a growing body of evidence that a single multifunctional protein phosphatase is responsible for the dephosphorylation of several of the enzymes involved in the regulation of glycogen metabolism (see Chapter 1). This enzyme, termed protein phosphatase-1, can dephosphorylate sites phosphorylated by various types of protein kinases including cyclic AMP-dependent and calcium-dependent protein kinases. However, dephosphorylation at all of these sites affects metabolism in the same direction, i.e., to promote glycogen synthesis or prevent glycogen breakdown. Burchell et al. [82] have recently obtained evidence that protein phosphatase-1 is present in various tissues, such as adipose tissue, brain and mammary gland, at levels similar to those present in skeletal muscle. Since the enzymes of glycogen metabolism are almost completely absent from these tissues [82], protein phosphatase-1 may play a general role in the regulation of intracellular metabolism.

I have already described how protein phosphatase-1 from skeletal muscle can activate and dephosphorylate purified rabbit mammary acetyl-CoA carboxylase (Section 4.2). This does not, of course, prove that protein phosphatase-1 is the enzyme responsible for the dephosphorylation of acetyl-CoA carboxylase in vivo. However, we have recently obtained evidence in favour of this latter hypothesis (J.G. Foulkes and D.G. Hardie, unpublished observations).

Extracts of rabbit skeletal muscle or lactating rabbit mammary gland contain protein phosphatases that are active against both phosphorylase a and acetyl-CoA carboxylase phosphorylated in site 1, and in addition the activity ratio phosphorylase phosphatase/acetyl-CoA carboxylase phosphatase is almost identical in both tissues. This is a revealing observation since acetyl-CoA carboxylase is scarcely detectable in skeletal muscle while phosphorylase is virtually absent from the mammary gland. We have used inhibitor-1 and -2 as probes to investigate the nature of these protein phosphatase activities. These inhibitors are heat-stable proteins from skeletal muscle which inhibit protein phosphatase-1 but no other protein phosphatase that has been tested (Chapter 1). The phosphatase activities against either phosphorylase a or acetyl-CoA carboxylase can be inhibited in parallel by inhibitor-1 or -2. A maximum of 80% inhibition was observed in muscle extracts and 60% inhibition in mammary gland extracts. These results suggest that protein phosphatase-1 is the major protein phosphatase acting on these two phosphoproteins in either muscle or mammary gland. In addition we have shown that the phosphorylase phosphatase and acetyl-CoA carboxylase

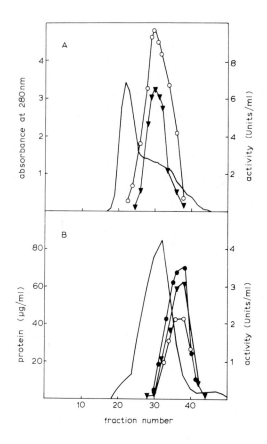

Figure 5. Purification of acetyl-CoA carboxylase phosphatase from lactating rabbit mammary gland. (A) Chromatography of a 35%-55% saturated ammonium sulphate fraction on Sephadex G-200. (B) Fractions containing acetyl-CoA carboxylase phosphatase activities from the gel filtration shown in (A) were pooled, precipitated with 80% ethanol at room temperature, and rechromatographed on the same Sephadex G-200 column. The fraction size is the same as in (A). Key: —— A_{280} or protein concentration; ▼——▼ phosphorylase phosphatase; ○——○ acetyl-CoA carboxylase phosphatase (site 1); ●——● acetyl-CoA carboxylase phosphatase (sites 1 + 2). (J.G. Foulkes and D.G. Hardie, unpublished work).

phosphatase activities from lactating mammary gland copurify through four purification steps. The activity against either substrate is purified 250-fold in over 30% yield. Particularly striking is the behaviour of the two activities on gel filtration. At step two, gel filtration on Sephadex G-200, both activities elute together with an apparent molecular weight of about 200 000 (Figure 5A). However after ethanol precipitation at room temperature (step 3), both activities elute from Sephadex G-200 with an apparent molecular weight of about 120 000 (Figure 5B). It is extremely unlikely that two distinct protein phosphatases would exhibit such unusual behaviour.

Although acetyl-CoA carboxylase phosphatase was purified using enzyme phosphorylated by acetyl-CoA carboxylase kinase-2 as substrate (site 1) the purified mammary phosphatase also dephosphorylated both sites on acetyl-CoA carboxylase phosphorylated by cyclic AMP-dependent protein kinase (sites 1 and 2).

5. Phosphorylation of fatty acid synthase

In 1975 Qureshi et al. [83] reported preliminary evidence that pigeon liver fatty acid synthase is regulated in vitro by a phosphorylation–dephosphorylation mechanism. Two forms of the enzyme which differed in specific activity could be separated by chromatography on Sepharose ϵ-aminocaproyl pantetheine. These two forms could apparently be interconverted using conditions expected to favour phosphorylation or dephosphorylation. Radioactivity from $[\gamma\text{-}^{32}P]ATP$ was incorporated into the low specific activity form, and was released on reconversion to the high specific activity form.

As yet no other laboratory has obtained any evidence for the phosphorylation of fatty acid synthase. Rabbit mammary fatty acid synthase purified in this laboratory does not contain significant quantities of alkali-labile phosphate, and is not phosphorylated by cyclic AMP-dependent protein kinase (Table 2), by endogenous protein kinases [30], or by phosphorylase kinase (D.G. Hardie and P. Cohen, unpublished observations). In addition, Brownsey et al. [75] have reported that fatty acid synthase is not phosphorylated in isolated rat fat pads. In view of these negative results, additional experiments are required before the phosphorylation of fatty acid synthase can be accepted. In particular, the original report [83] did not comment on the stoichiometry of phosphorylation, and it was not shown that the ^{32}P-radioactivity associated with the protein was bound covalently. Until such time as these experiments are carried out, the phosphorylation of fatty acid synthase must be regarded as not proven.

6. Summary and perspectives

In this final section, I would like to draw together several of the themes that have emerged from my discussion and also consider the direction in which future research may go.

There can be few who would now doubt that reversible phosphorylation is an important mechanism for the regulation of acetyl-CoA carboxylase. Phosphorylation has been demonstrated both with the purified enzyme (Section 4.2) and in vivo (Section 4.3). It has also been shown that the kinetic properties of the purified enzyme change as a consequence of changes in the phosphorylation state. Similarly, treatment with hormones either in vivo or with intact cell preparations produces parallel changes in the kinetic properties and the state of phosphorylation of the enzyme. The time now seems right to combine the information obtained with the purified enzyme and with intact cell preparations in an attempt to show that hormones regulate fatty acid synthesis via changes in the phosphorylation of specific sites on acetyl-CoA carboxylase. However this analysis is complicated by the finding that acetyl-CoA carboxylase is phosphorylated by multiple protein kinases at multiple sites. It will be necessary to fully characterise each phosphorylation site on the enzyme to determine the effect of phosphorylation on the kinetic properties and the state of polymerisation of the enzyme. Further purification of the various protein kinases and phosphatases is also required in order that their regulation and their phosphorylation site specificities can be studied. It will then be possible to compare the sites phosphorylated in vitro with the sites phosphorylated in vivo in response to hormones.

At this stage, it is instructive to compare the regulation of glycogen synthesis with that of fatty acid synthesis. This comparison is useful because our understanding of the regulation of glycogen synthesis is at a more advanced stage and can be used as a model with which to frame testable hypotheses about the regulation of fatty acid synthesis. From the point of view of cellular economy, it would perhaps be surprising if these highly analogous pathways were regulated by radically different mechanisms.

The analogies between the regulation of the two pathways are already quite striking. Glycogen synthesis and fatty acid synthesis are affected rapidly and in parallel by several hormones, including glucagon, adrenaline (β-adrenergic action), insulin and vasopressin (Section 2). α-adrenergic agents, which bring about an inhibition of glycogen synthesis [20], do not yet appear to have been tested on fatty acid synthesis. The two pathways are regulated in the short-term by alterations in the activity of the rate-limiting enzymes, glycogen synthase and acetyl-CoA carboxylase. These two enzymes are subject to allosteric regulation, both being activated by precursors of the pathway which they regulate, i.e., glucose-6-phosphate and citrate. Both enzymes are also subject to covalent modification, and are phosphorylated at multiple sites by both cyclic AMP-dependent and -independent protein kinases.

The interconversions of the various phosphorylated forms of glycogen

58

synthase and acetyl-CoA carboxylase are summarised in Figures 6 and 7. Glycogen synthase is phosphorylated by at least three different protein kinase at distinct sites, these protein kinases being as follows:

   (1) Cyclic AMP-dependent protein kinase, which mediates the action of glucagon and β-adrenergic agents on glycogen synthesis.
   (2) Glycogen synthase kinase-2, now known to be phosphorylase kinase [84,85]. Phosphorylation by this calcium-dependent protein kinase synchronises the inhibition of glycogen synthesis with the onset of muscle contraction, and may also mediate the effects of α-adrenergic agents in the liver [86].
   (3) Glycogen synthase kinase-3. The regulation and function of this protein kinase is not understood, although one possibility is that it mediates the effects of insulin.

Phosphorylation by all of these protein kinases can be reversed by a single protein phosphatase, protein phosphatase-1. The dephosphorylation reactions are also regulated by cyclic AMP: protein phosphatase-1 is inhibited by a heat-stable inhibitor protein, termed inhibitor-1, which is only active after phosphorylation by cyclic AMP-dependent protein kinase.

The reader is referred to the introductory chapter in this volume for a more detailed review of glycogen metabolism. If we now compare the known inter-conversion reactions of acetyl-CoA carboxylase (Figure 7) the similarities are striking. Acetyl-CoA carboxylase is phosphorylated by cyclic AMP-dependent protein kinase and by at least one cyclic AMP-independent protein kinase, acetyl-CoA carboxylase kinase-2. Both of these phosphorylations can be reversed by protein phosphatase-1 (Section 4.4).

Using the regulation of glycogen synthesis as a model, several testable

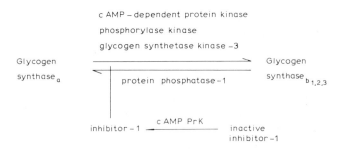

Figure 6. Summary of the enzymes that have been shown to affect the state of phosphorylation of skeletal muscle glycogen synthase in vitro. Glycogen synthase a represents the active dephosphorylated form and glycogen synthase $b_{1,2,3}$ the forms phosphorylated by cyclic AMP-dependent protein kinase (cAMP PrK), phosphorylase kinase and glycogen synthase kinase-3 respectively.

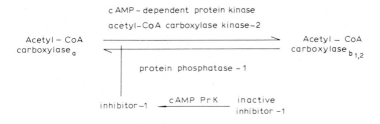

Figure 7. Summary of the enzymes that have been shown to affect the phosphorylation state of rabbit mammary acetyl-CoA carboxylase in vitro. Acetyl-CoA carboxylase$_a$ represents the active dephosphorylated form and acetyl-CoA carboxylase$_{b1,2}$ the less active forms phosphorylated by cyclic AMP-dependent protein kinase (cAMP PrK) and acetyl-CoA carboxylase-2 respectively.

hypotheses can be formed about the regulation of acetyl-CoA carboxylase. These are most conveniently presented as a list of questions:

(1) Do α-adrenergic agents affect fatty acid synthesis?

(2) Do insulin, α-adrenergic agents, vasopressin or angiotensin II affect the state of phosphorylation of acetyl-CoA carboxylase in vivo?

(3) Is there a calcium-dependent protein kinase which acts on acetyl-CoA carboxylase?

(4) What is the mechanism of regulation of acetyl-CoA carboxylase kinase-2?

(5) Is acetyl-CoA carboxylase kinase-2 related to any of the cyclic AMP-independent protein kinases acting on glycogen synthase?

(6) Can hormones which increase the intracellular concentration of cyclic AMP inhibit the dephosphorylation of acetyl-CoA carboxylase through the phosphorylation of inhibitor-1?

These questions should provide fruitful areas of investigation in the future.

It is worth pointing out that the regulation of two different metabolic pathways by common protein kinases and phosphatases need not imply that the overall rates of the pathways are inextricably linked. Superimposed on the covalent regulation there is, of course, an additional tier of allosteric regulation. These two levels of regulation are interdependent in two different ways. Firstly changes in the state of phosphorylation can modulate the response to the allosteric effector. All the phosphorylated forms of acetyl-CoA carboxylase so far described are activated by physiological concentrations of citrate, but the magnitude of the effect depends on the state of phosphorylation. Secondly, the concentration of allosteric effectors can affect the rate of phosphorylation or dephosphorylation. There is already some evidence that high concentrations of citrate inhibit the phosphorylation of acetyl-CoA carboxylase [62].

These finer points of the regulation of acetyl-CoA carboxylase await the future experimenter. It is a reflection on science in general that the discovery of a new system of regulation by reversible phosphorylation poses more new questions than it answers.

References

1. Romsos, D.R. and Leveille, G.A. (1974) Advan. Lipid Res., 12, 97–146.
2. Bloch, K. and Vance, D. (1977) Ann. Rev. Biochem., 46, 263–298.
3. Klain, G.J. and Weiser, P.C. (1973) Biochem. Biophys. Res. Commun., 55, 76–83.
4. Cook, G.A., Nielsen, R.C., Hawkins, R.A., Mehlman, M.A., Lakshmanan, M.R. and Veech, R.L. (1977) J. Biol. Chem., 252, 4421–4424.
5. Watkins, P.A., Tarlow, D.M. and Lane, M.D. (1977) Proc. Natl. Acad. Sci. USA, 74, 1497–1501.
6. Geelen, M.J.H., Beynen, A.C., Christiansen, R.Z., Lepreau-Jose, M.J. and Gibson, D.M. (1978) FEBS Lett., 95, 326–330.
7. Witters, L.A., Kowaloff, E.M. and Avruch, J. (1979) J. Biol. Chem., 254, 245–248.
8. Ma, G., Gove, C.D. and Hems, D.A. (1977) Biochem. Soc. Trans., 5, 986–989.
9. Lee, K-H. and Kim, K.H. (1979) J. Biol. Chem., 254, 1450–1453.
10. Lee, K-H. and Kim, K.H. (1978) J. Biol. Chem., 253, 8157–8161.
11. Brownsey, R.W., Hughes, W.A and Denton, R.M. (1979) Biochem. J., 184, 23–32.
12. Stansbie, D., Brownsey, R.W., Crettaz, M. and Denton, R.M. (1976) Biochem. J., 160, 413–416.
13. McCormack, J.G. and Denton, R.M. (1977) Biochem. J., 166, 627–630.
14. Lee, K-H., Thrall, T. and Kim, K.H. (1973) Biochem. Biophys. Res. Commun., 54, 1133–1140.
15. Halestrap, A.P. and Denton, R.M. (1973) Biochem. J., 132, 509–517.
16. Ma, G.Y. and Hems, D.A. (1975) Biochem. J., 152, 389–392.
17. Hems, D.A. (1977) FEBS Lett., 80, 237–245.
18. Bishop, J.S. and Larner, J. (1967) J. Biol. Chem., 242, 1354–1356.
19. Craig, J.W. and Larner, J. (1964) Nature, 202, 971–973.
20. Hutson, N.J., Brumley, F.T., Assimacopoulos, F.D., Harper, S.C. and Exton, J.H. (1976) J. Biol. Chem., 251, 5200–5208.
21. Hems, D.A., Rodrigues, L.M. and Whitton, P.D. (1976) Biochem. J., 160, 367–374.
22. Villar-Palasi, C. and Larner, J. (1960) Biochim. Biophys. Acta, 39, 171–173.
23. Roach, P.J., Takeda, Y. and Larner, J. (1976) J. Biol. Chem., 251, 1913–1919.
24. Gregolin, C., Ryder, E. and Lane, M.D. (1968) J. Biol. Chem., 243, 4227–4235.
25. Nakanishi, S. and Numa, S. (1970) Eur. J. Biochem., 16, 161–173.
26. Inoue, H. and Lowenstein, J.M. (1973) J. Biol. Chem., 247, 4825–4832.
27. Moss, J., Yamagishi, M., Kleinschmidt, A.K. and Lane, M.D. (1972) Biochemistry, 11, 3779–3786.
28. Ahmad, F., Ahmad, P.M., Pieretti, L. and Watters, G.T. (1978) J. Biol. Chem., 253, 1733–1737.
29. Manning, R., Dils, R. and Mayer, R.J. (1976) Biochem. J., 153, 463–468.
30. Hardie, D.G. and Cohen, P. (1978) Eur. J. Biochem., 92, 25–34.
31. Gregolin, C., Ryder, E., Kleinschmidt, A.K., Warner, R.C. and Lane, M.D. (1966) Proc. Natl. Acad. Sci. USA, 56, 148–155.
32. Henninger, G. and Numa, S. (1972) Hoppe-Seyler's Z. Physiol. Chem., 353, 459–462.
33. Moss, J. and Lane, M.D. (1972) J. Biol. Chem., 247, 4944–4951.
34. Guchhait, R.B., Polakis, S.E., Dimroth, P., Stoll, E., Moss, J. and Lane, M.D. (1974) J. Biol. Chem., 249, 6633–6645.

35. Matsuhashi, M., Matsuhashi, S. and Lynen, F. (1963) Biochem. Z., 340, 263–289.
36. Guchhait, R.B., Zwergel, E. and Lane, M.D. (1974) J. Biol. Chem., 249, 4776–4780.
37. Tanabe, T., Wada, K., Okazaki, T., Numa S. (1975) Eur. J. Biochem., 57, 15–24.
38. Mackall, J. and Lane, M.D. (1977) Biochem. J., 162, 635–642.
39. Brownsey, R.W., Hughes, W.A., Denton, R.M. and Mayer, R.J. (1977) Biochem. J., 168, 441–445.
40. Kirschner, K. and Bisswanger, H. (1976) Ann. Rev. Biochem., 45, 143–166.
41. Gregolin, C., Ryder, E., Warner, R.C., Kleinschmidt, A.K., and Lane, M.D. (1966) Proc. Natl. Acad. Sci. USA, 56, 1751–1758.
42. Gregolin, C., Ryder, E., Warner, R.C., Kleinschmidt, A.K., Chang, H.C. and Lane, M.D. (1968) J. Biol. Chem., 243, 4236–4245.
43. Meredith, M.J. and Lane, M.D. (1978) J. Biol. Chem., 253, 3381–3383.
44. Brady, R.O. and Gurin, S. (1952) J. Biol. Chem., 199, 421–431.
45. Martin, D.B. and Vagelos, P.R. (1962) Biochem. Biophys. Res. Commun., 7, 101–106.
46. Vagelos, P.R., Alberts, A.W. and Martin, D.B. (1962) Biochem. Biophys. Res. Commun., 8, 4–8.
47. Moss, J. and Lane, M.D. (1972) J. Biol. Chem., 247, 4944–4951.
48. Numa, S., Ringelmann, E. and Lynen, F. (1965) Biochem. Z., 343, 243–257.
49. Lane, M.D., Moss, J. and Polakis, S.E. (1975) in Subunit Enzymes – Biochemistry and Function (ed. K.E. Ebner), pp. 181–221, Marcel Dekker, New York.
50. Siess, E.A., Brocks, D.G. and Wieland, O.H. (1978) Hoppe-Seyler's Z. Physiol. Chem., 359, 785–798.
51. Brownsey, R.W., Bridges, B.J. and Denton, R.M. (1977) Biochem. Soc. Trans., 5, 1286–1288.
52. Bortz, W.M. and Lynen, F. (1963) Biochem. Z., 337, 505–509.
53. Taketa, K. and Pogell, B.M. (1966) J. Biol. Chem., 241, 720–726.
54. Zahler, W.L., Barden, R.E. and Cleland, W.W. (1968) Biochim. Biophys. Acta, 164, 1–11.
55. Goodridge, A.G. (1972) J. Biol. Chem., 247, 6946–6952.
56. Ogiwara, H., Tanabe, T., Nikawa, J. and Numa, S. (1978) Eur. J. Biochem., 89, 33–41.
57. Carlson, C.A. and Kim, K.H. (1973) J. Biol. Chem., 248, 378–380.
58. Carlson, C.A. and Kim, K.H. (1974) Arch. Biochem. Biophys., 164, 478–489.
59. Carlson, C.A. and Kim, K.H. (1974) Arch. Biochem. Biophys., 164, 490–501.
60. Halestrap, A.P. and Denton, R.M. (1974) Biochem. J., 142, 365–377.
61. Lane, M.D., Moss, J. and Polakis, S.E. (1974) Curr. Top. Cell. Regul., 8, 139–195.
62. Lent, B.A., Lee, K-H. and Kim, K.H. (1978) J. Biol. Chem., 253, 8149–8156.
63. Smith, A. and Williamson, I.P. (1977) Biochem. Soc. Trans., 5, 737–739.
64. Desjardins, P.R. and Dakshinamurti, K. (1978) Int. J. Biochem., 9, 227–234.
65. Walker, J.H., Betts, S.A., Manning, R. and Mayer, R.J. (1976) Biochem. J. 159, 355–362.
66. Lee, K-H. and Kim, K.H. (1977) J. Biol. Chem., 252, 1748–1751.
67. Beavo, J.A., Bechtel, P.J. and Krebs, E.G. (1974) Proc. Natl. Acad. Sci. USA, 71, 3580–3583.
68. Hardie, D.G. and Cohen, P. (1978) FEBS Lett., 91, 1–7.
69. Hardie, D.G. and Cohen, P. (1979) FEBS Lett., 103, 333–338.
70. Reimann, E.M., Brostrom, C.O., Corbin, J.D., King, C.A. and Krebs, E.G. (1971) Biochem. Biophys. Res. Commun., 42, 187–194.
71. Walsh, D.A., Perkins, J.P. and Krebs, E.G. (1968) J. Biol. Chem., 243, 3763–3765.
72. Nimmo, H.G. and Cohen, P. (1974) FEBS Lett., 47, 162–166.
73. Walsh, D.A., Perkins, J.P., Brostrom, C.O., Ho, E.S. and Krebs, E.G. (1971) J. Biol. Chem., 246, 1968–1976.
74. Cohen, P. (1978) Curr. Top. Cell. Regul., 14, 117–196.
75. Bylund, D.B. and Krebs, E.G. (1975) J. Biol. Chem., 250, 6355–6361.
76. Brownsey, R.W., Hughes, W.A., Denton, R.M. and Mayer, R.J. (1977) Biochem. J., 168, 441–445.

77. Avruch, J., Witters, L.A., Alexander, M.C. and Bush, M.A. (1978) J. Biol. Chem., 253, 4753–4761.
78. Witters, L.A., Kowaloff, E.M. and Avruch, J. (1979) J. Biol. Chem., 254, 245–248.
79. Pekala, P.H., Meredith, M.J., Tarlow, D.M. and Lane, M.D. (1978) J. Biol. Chem., 253, 5267–5269.
80. Lee, K-H. and Kim, K.H. (1979) J. Biol Chem., 254, 1450–1453.
81. Brownsey, R.W., Hughes, W.A. and Denton, R.M. (1979) Biochem. J., 184, 23–32.
82. Burchell, A., Foulkes, J.G., Cohen, P.T.W., Condon, G.D. and Cohen, P. (1978) FEBS Lett., 92, 68–72.
83. Qureshi, A.A., Jenik, R.A., Kim, M., Lornitzo, F.A. and Porter, J.W. (1975) Biochem. Biophys. Res. Commun., 66, 344–351.
84. Roach, P.J., DePaoli-Roach, A.A. and Larner, J. (1978) J. Cyc. Nuc. Res., 4, 245–257.
85. Embi, N., Rylatt, D.B. and Cohen, P. (1979) Eur. J. Biochem., 100, 339–347.
86. Assimacopoulos-Jeannet, F.D., Blackmore, P.F. and Exton, J.H. (1977) J. Biol. Chem., 252, 2662–2669.

CHAPTER 4

Reversible phosphorylation of hydroxymethylglutaryl CoA reductase

THOMAS S. INGEBRITSEN AND DAVID M. GIBSON

1. Introduction

Cholesterol synthesis in mammalian cells is regulated principally through the enzyme HMG-CoA reductase (hydroxymethylglutaryl CoA reductase (NADPH), EC 1.1.1.34*). In most mammalian cells the enzyme is tightly bound to the endoplasmic reticulum [9]. Although the enzyme protein cannot be removed from the microsomal fraction by conventional washing techniques, it is solubilized after the membranes are subjected to slow freezing and thawing [10,11]. Reductase can also be extracted into phosphate buffer supplemented with 4 M KCl without the need for freezing and thawing the microsomal preparation [11]. Since none of these techniques grossly alters the bilaminar structure of the microsomal membranes [11], reductase appears to be a 'peripheral' rather than an 'integral' membrane protein according to the definition of Singer and Nicolson [12].

Solubilized reductase and the microsomal-bound enzyme share similar kinetic properties in regard to the K_m values for HMG-CoA and NADPH [4]. The solubilized enzyme, however, differs from the membrane bound reductase, in that the former is cold inactivated. Depending on the isolation technique, the loss of activity at 4°C is either irreversible [11], or is readily reactivated by incubation at 37°C [13]. The irreversible, but not the reversible, loss of activity is precluded if 4 M KCl is added to the storage buffer [11]. Tormanen et al. [14] have recently suggested that the irreversible cold lability is not an intrinsic property of reductase, since a more highly purified preparation of the enzyme was no longer subject to inactivation at 4°C. The susceptibility of reductase to cold inactivation is reminiscent of a

---

* HMG-CoA reductase is simply designated 'reductase' in this article. The properties of reductase and its regulation have been extensively reviewed in the recent literature [1–8].

Cohen (ed.) Recently discovered systems of enzyme regulation by reversible phosphorylation
© Elsevier/North-Holland Biomedical Press, 1980

similar property of the Fl-ATPase solubilized from mitochondrial inner membrane. This characteristic was explained in terms of a conformational change in the Fl-ATPase attending solubilization of the enzyme [15].

Solubilized reductase has now been purified to homogeneity by several groups [16–19]. The purified enzyme has a molecular weight of 200 000 and appears to be composed of four identical 50 000 dalton subunits [2].

Liver reductase has been the focus of many investigations since hepatic cholesterol formation is geared not only to replacement of liver cell membrane components [20,21] but also for net synthesis of the plasma lipoproteins [22] and the bile acids [23]. Normally most extrahepatic cells synthesize relatively little cholesterol (per gram of tissue) even though they possess the potential for much greater biosynthetic activity. De novo cholesterol synthesis in these tissues is held in abeyance by the release of free cholesterol from plasma low density lipoproteins (LDL) that are brought into cells by endocytosis [24]. Hepatic cholesterol biosynthesis is also subject to regulation by exogenous (dietary) cholesterol delivered to the cell by chylomicron remnants [25].

The biochemical mechanism(s) for this feedback regulation of cholesterol biosynthesis at the HMG-CoA reductase step has been studied extensively. 'Long term' induction and repression of enzyme synthesis is thought to play an important role [26–28]. This mechanism is facilitated by the very short half life of reductase in liver and in cultured cells (on the order of 2–4 h) [28–33]. Nevertheless, evidence is discussed in this review indicating that 'short term' control effected by other mechanisms may also be important [17, 26, 34–36].

Reductase activity in rodent livers undergoes a pronounced diurnal variation [26,29,32,37–46]. In rats maintained on a controlled lighting schedule (in which lights are on for 12 h per day) enzyme activity rises to a maximal level (10-fold over the minimum) 6 h into the dark phase of the cycle while the activity falls to its nadir at 6 h into the light phase [29, 38–42, 45, 46]. These changes in reductase activity while occurring synchronously with the lighting cycle are keyed to the nocturnal feeding behavior of rats. After training rats to feed (in a 2-h period) during the light phase of the cycle, peak reductase activity occurs at 4–6 h after feeding commences [29]. A considerable body of evidence supports the concept that changes in reductase activity during this cycle are associated with changes in the rate of enzyme synthesis [29,32,33,37,43,45,46].

Porter's laboratory has shown that insulin plays an important role in maintaining the normal diurnal rhythm of reductase in rats [47–50]. When rats are made diabetic with streptozotocin, reductase activity at the diurnal peak declines over a 7-day period to a level that is about half of the normal diurnal minimum. Reductase activity at the nadir of the cycle similarly declines to this low level. Daily injection of the diabetic rats with a slow-acting form of insulin (i.e., protamine zinc insulin) restores the normal activities at the peak and nadir of the cycle within 4 days. Diabetic rats administered a fast-acting form of insulin (regular insulin) 2 h prior to sacrifice experience a 10-fold increase in activity at the time of the normal diurnal maximum to a value nearly identical to that attained by normal rats at this time of day. Injection of regular insulin prior to the nadir of the cycle resulted in only a 2-fold increase in reductase activity again to a value comparable to that attained by normal animals during this period [48]. These data suggest that insulin plays a permissive role in which the presence of the hormone is required to maintain the normal periodic fluctuation of reductase activity.

In addition to the permissive role, there is evidence to suggest that insulin may actually signal the normal diurnal rise in reductase activity. Treatment of either normal or 2-day diabetic rats with insulin during the period of diurnal minimum produces a rapid 3- to 8-fold rise in reductase activity (maximum in 2 h) to a value well above the normal diurnal minimum in both cases [47, 51, 52]. This rapid increase is blunted, however, in long-term diabetic rats (see above).

Administration of glucagon or cAMP markedly diminished the normal diurnal rise in reductase activity [48, 49]. Furthermore, glucagon was found to prevent the insulin stimulation of reductase activity in either normal or diabetic rats [47,49].

Insulin and glucagon effects have been observed in a variety of in vitro systems. Gibson and Geelen found that insulin stimulated both acetate and tritiated water incorporation into the non-saponifiable lipid fraction (principally cholesterol) during 1–3-h incubations with isolated hepatocytes [53,54]. Bhathena et al. [55] noted that insulin at levels as low as 7 nM markedly stimulated acetate incorporation into non-saponifiable lipids and fatty acids and the activity of HMG-CoA reductase in human fibroblasts, rabbit aortic cells, and human lung WI-38 cells.

Geelen and Gibson [53,54] observed that glucagon at concentrations as low as 10^{-9} M inhibited acetate incorporation into cholesterol in hepatocytes and also blocked the insulin-induced increase in cholesterogenesis in these cells. Ma et al. [56] in perfused mouse liver have shown that glucagon in the concentration range of $10^{-10}–10^{-9}$ M produced a potent inhibition of tritiated water incorporation into cholesterol. That glucagon acts at the level of HMG-CoA reductase in hepatocytes has been confirmed by Edwards and co-workers [57]. This glucagon effect appears to be mediated via cAMP as a second messenger since many studies with liver slices [58] and isolated hepatocytes have shown that cAMP diminishes both cholesterol synthesis and the activity of HMG-CoA reductase [59–62].

In terms of short-term regulation of liver metabolism insulin and glucagon are known to signal a pervasive dephosphorylation and phosphorylation, respectively, of a set of key regulatory enzymes including glycogen phosphorylase, phosphorylase kinase, glycogen synthase, pyruvate kinase, pyruvate dehydrogenase and acetyl-CoA carboxylase [63]. Recent evidence, reviewed here, suggests that HMG-CoA reductase should be added to this list.

2. Short-term modulation of reductase

Since signal transduction, subsequent to the binding of glucagon to the surface of the hepatocyte, is directed through cAMP-dependent protein kinases [64] it is possible that the reductase enzyme is affected by short-term, reversible phosphorylation. Pursuing this point Beg et al. in 1973 [59] demonstrated that reductase activity in isolated rat liver microsomes was severely diminished if this preparation was preincubated in the presence of millimolar levels of both ATP and $MgCl_2$ for 10–20 min at 37°C (Table 1). The microsomes so treated could be spun out of the incubation medium (100 000 × g for 60 min and resuspended in an $ATP(Mg^{2+})$-free solution without changing the depressed state of the enzyme. Subsequent addition of protein fractions from the liver cytosol restored activity in a time- and concentration-dependent manner [59].

After serial extraction of freshly-isolated microsomes under conditions that did not remove the tightly-bound reductase there was a progressive loss of the capacity to respond to ATP(Mg) inhibition ([59]; Table 1). Depleted microsomes could be restored to the initial state by adding back the microsomal extracts, liver cytosol (Table 1), or the protein fraction of the cytosol precipitating at 30% saturation with ammonium sulfate [59,65,66]. Similarly, reductase, solubilized by freezing and thawing microsomes according to the method of Brown et al. [11], was inhibited in the presence of ATP(Mg) and this cytosolic protein fraction [65]. The accumulated evidence in these early studies was consistent with the concept that reductase could be controlled by reversible, covalent modulation [59].

A number of laboratories corroborated these initial findings. Goodwin and Margolis independently observed that the rate of acetate incorporation into cholesterol by a 10 000 × g supernatant from rat liver homogenates was enhanced 7- to 20-fold following preincubation for up to 20 min at 37°C [67–69]. Synthesis of cholesterol from mevalonate was not affected by the preincubation. Acetate incorporation into CO_2 and fatty acids was also enhanced but to a much smaller extent (2–3-fold vs 7–20-fold). These studies

TABLE 1

HMG-CoA reductase activity in washed microsomes

Microsomes	Additions	Reductase activity (nmol/min/mg)
(A)		
Unwashed (0.81 mg)	0	0.31 ± 0.02
	ATP (Mg^{2+})	0.08 ± 0.01
Washed 1 × (0.67 mg)	0	0.39 ± 0.01
	ATP (Mg^{2+})	0.20 ± 0.02
Washed 2 × (0.66 mg)	0	0.38 ± 0.06
	ATP (Mg^{2+})	0.35 ± 0.03
Washed 3 × (0.65 mg)	0	0.38 ± 0.02
	ATP (Mg^{2+})	0.39 ± 0.01
(B)		
Washed 3 ×	0	0.28 ± 0.01
Washed 3 × + cytosol	0	0.31 ± 0.02
Washed 3 × + cytosol	ATP (Mg^{2+})	0.03 ± 0.01

In experiment A microsomes were preincubated with and without ATP and Mg^{2+} (2 mM each) for 15 min, following which reductase activity was measured. In experiment B thrice-washed microsomal pellets (0.26 mg) were preincubated, as indicated, with cytosol (0.33 mg) and with ATP (1 mM) and Mg^{2+} (1 mM) for 10 min at 37 °C. After this period reductase activity was determined [59].

indicated that a step between acetyl-CoA and mevalonic acid was the site of activation. Later examination of this phenomenon [70,71] in which microsomes were isolated from the 10 000 × g supernatant before and after incubation at 37 °C demonstrated that microsomal HMG-CoA reductase was activated. Enhancement of both cholesterol synthesis and HMG-CoA reductase activity required the simultaneous presence of both cytosol and microsomes during the preincubation period, indicating that the activating enzyme was of cytosolic origin. Interestingly, it was found that both reductase activation and enhancement of cholesterol synthesis were prevented when cAMP (10^{-6}–10^{-3} M) and MgATP were included in the incubation. MgATP alone was without effect.

Berndt et al. [72,73] reported a similar activation of HMG-CoA reductase in mouse liver homogenates and subfractions. Activation of reductase was time- and temperature-dependent (blocked at low temperature) and required the presence of the cytosol. Incubation of the cytosol with trypsin prevented the activation. Of considerable import was the observation that activation was prevented by 100 mM NaF, a known inhibitor of phosphoprotein phosphatases.

Berndt and co-workers in a subsequent paper [73] noted that the cytosolic activating system could be replaced with a mixture of Na_2SO_3 (but not Na_2SO_4) and $MgCl_2$. Furthermore, it was found that this kind of activation was blocked by NaF. This approach derived from an earlier study by Hizukuri and Larner [74] who showed that glycogen-bound glycogen synthase *b* could be activated (dephosphorylated) by incubation with a similar mixture. By inference, the activation of reductase could result from dephosphorylation of the reductase enzyme. The recent observation by Shimazu and co-workers [75] that phosphorylase phosphatase is severely inhibited by oxidized glutathione may explain the mechanism for activation of the reductase by this mixture. Since in the work of Berndt et al. [73] no thiol protecting reagent was included ·in the medium employed for isolation of the microsomes, incubation of microsomes with SO_3^{-2} could have activated endogenous (inhibited) microsomal phosphatase which then catalyzed dephosphorylation of the reductase.

Brown et al. [76] extended the early work of Beg et al. [59] with the finding that microsomal HMG-CoA reductase in fibroblast extracts was subject to inactivation with MgATP. The inactivating factor in fibroblast extracts also brought about a severe impairment of rat liver microsomal reductase activity.

·3. Properties of inactive reductase

Several lines of evidence indicate that a stable, inactive form of reductase is the primary product resulting from incubation with MgATP. First, Nordstrom et al. [77] demonstrated that pretreatment of microsomes with MgATP while inactivating reductase, did not alter the recovery of a tracer amount of mevalonic acid added at the beginning of the reductase assay. Consequently, the diminished rate of formation of mevalonic acid by inactivated microsomal reductase did not result from removal of mevalonic acid by the subsequent enzyme in the cholesterol biosynthetic pathway (mevalonate kinase). Secondly, the reduction in reductase activity was not due to the activation of an HMG-CoA utilizing enzyme such as HMG-CoA lyase or HMG-CoA deacylase since the K_m for HMG-CoA of the inactive enzyme was the same as that for the active form [76,77] (see below). Thirdly, the fact that partially purified soluble reductase is subject to inactivation by MgATP [77] suggested that the microsomal membrane is not absolutely needed to effect the inactivation, and further ruled out the possibility that inactivation of the reductase is primarily due to an alteration in the properties of the microsomal membrane. Finally, the idea that either low or high molecular weight inhibitors of reductase were generated during preincubation with MgATP became untenable, viz. inactive reductase was not restored by treatments such as dialysis [77], reisolation of the microsomes [59], or desalting over Sephadex G-25 [78] which remove small molecules such as Mg^{2+}, ATP, or derivative nucleotides. Brown et al. [76] incubated microsomes and a fibroblast extract with ATP and Mg^{2+} and then after addition of EDTA they tested the complete mixture, microsomes isolated from the mixture, and the soluble fraction from the mixture to see if any of these fractions would inhibit reductase in freshly isolated microsomes. None did. Ingebritsen [78] incubated unwashed microsomes with MgATP and then attempted to extract any inhibitor after treatment with perchloric acid. These extracts when added back to fresh microsomes neither inhibited reductase nor enhanced the inactivation of reductase by MgATP. Taken together, these observations indicated that incubation of microsomal reductase with MgATP produces a stable alteration in the reductase protein molecule itself.

The reduced activity of reductase observed after treatment with MgATP could result either from a change in the V_{max} or in the affinity of the enzyme for the two substrates, HMG-CoA and NADPH. Brown et al. [76] and Nordstrom et al. [77] independently determined that the K_m for HMG-CoA was the same for both forms of the enzyme. Brown et al. [76] further observed that the K_m for NADPH was not altered by inactivation. By contrast the V_{max} reductase fell from 323 pmol·min^{-1}·mg^{-1} to 26 pmol·min^{-1}·mg^{-1} during maximal inactivation [77].

Although incubation of microsomes with MgATP or MgATP supplemented with factors derived from the cytosol, microsomal extracts or fibroblast extracts resulted in a prompt inactivation of the enzyme, the inactivation was never complete [59,76,77]. Brown et al. [76] studied this phenomenon in some detail using rat liver microsomal reductase. They found that the incomplete inactivation was not due to depletion of either MgATP or the inactivating factor during the incubation. Rather it appeared to represent a maximal state of inactivation of the enzyme. Two explanations are evident: (1) the inactive form of the reductase possesses a finite but much smaller activity than the active species or (2) two pools of reductase exist, one which is capable of being inactivated completely and another which is resistant to inactivation. The latter situation might result if a small amount of reductase were present on the inside surface of microsomal vesicles.

4. The reductase activating system

The extensive study of Nordstrom et al. in 1977 [77] provided decisive evidence that two enzymes were required for the interconversion of reductase. These were labelled the reductase inactivating and activating enzymes.

The activating system was purified from rat liver cytosol with acetone (35% to 60% fraction), ammonium sulfate (precipitation at 40% saturation), DEAE-cellulose and Sephadex G-150 [77]. Even though a 1300-fold purification had been achieved more than a dozen bands could be detected on polyacrylamide gel electrophoresis. The predominant peak on Sephadex G-150 in relation to known protein standards lay in a molecular weight class of 30 000.

Depressed reductase activity in ATP(Mg)-treated microsomes in the presence of EDTA-containing buffers was restored to full catalytic potential in a time-dependent manner with the purified activating enzyme. Fluoride at 50 mM completely blocked the process. Half-maximal inhibitory concentrations of fluoride were 5.0 mM or less. Homogenizing media containing fluoride were also used to obtain microsomal reductase from liver in its presumed in vivo state (see later).

Ingebritsen et al. [66] have provided evidence that the activating system may be replaced with liver phosphoprotein phosphatase. The striking similarity of the 35 000 dalton phosphorylase a phosphatase described by Brandt et al. [79–81] to the 30 000 dalton activating enzyme of Nordstrom et al. [77], led Ingebritsen and co-workers to test the possibility that partially purified liver phosphorylase phosphatase might reactivate the reductase. As seen in Figure 1, reductase was reactivated in a time- and dose-dependent manner by the phosphatase. Phosphorylase phosphatase and the activating activity copurified in fractions derived from the rat liver cytosol. The unique property of the phosphatase to be activated 15-fold by ethanol at 25 °C was shared by the reductase activating enzyme [1]. Both activities co-purified through two additional chromatographic separations on Sephadex G-75 and DEAE-Sephadex (Figure 2; [66]).

Further support for the concept that the reductase activating enzyme was identical to phosphorylase phosphatase was provided using specific phosphatase inhibitors [78]. NaF was found to inhibit both phosphorylase phosphatase and the reductase activating activities of the purified phosphatase in a parallel fashion. Half-maximal inhibition of both activities was attained at a fluoride concentration of 3–4 mM. Two heat stable proteins which are specific inhibitors of phosphorylase phosphatase [82] have been isolated from liver and skeletal muscle [82–86]. One of the proteins,

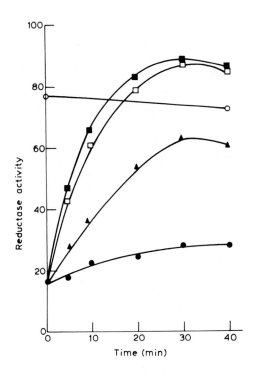

Figure 1. Reactivation of (Mg)ATP-treated microsomal reductase. Increasing amounts of liver cytosolic phosphorylase phosphatase separated on DEAE-Sephadex were added to inactivated microsomal reductase (159 μg microsomal protein) in imidazole buffer supplemented with 25 mM EDTA (100 μl final volume): 0 μg (●), 0.28 μg (▲), 0.70 μg (□) and 0.98 μg (■). The top horizontal line (○) is untreated microsomal reductase (169 μg protein per 100 μl) which was not fully activated as isolated from liver [77]. The ordinate is in milliunits of reductase activity, as a function of the preincubation time. The specific enzyme activity of this preparation was calculated to be 5.2 units of reductase reactivation activity per mg protein (phosphorylase phosphatase: 36.6 units/mg) [66].

designated as inhibitor 1, is only active after phosphorylation by cAMP-dependent protein kinase while the other is spontaneously active. Ingebritsen and Gibson [78] purified the spontaneously active inhibitor from rat liver by the method of Khandelwal and Zinman [86]. The purified inhibitor was found to decrease both phosphorylase phosphatase and the reductase activating activity in parallel (Figure 3). Similarly, when purified inhibitors 1 and 2 from skeletal muscle were used, both proteins were found to produce parallel inhibition of the phosphorylase phosphatase and reductase activating activities [78]. These studies conclusively showed that a dephosphorylation event was associated with the reductase activation. Since no evidence for the

72

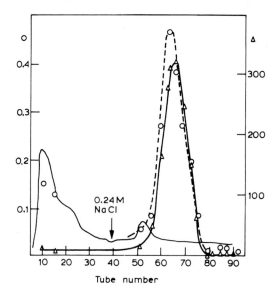

Figure 2. DEAE-Sephadex chromatography: 36 ml of a post ethanol fraction (per ml: 25.2 mg protein, 6.3 units of phosphorylase phosphatase and 1773 milliunits of reductase reactivation activity) were added to a column (30 × 2.5 cm) of DEAE-Sephadex A-50 which was previously equilibrated with 0.18 M NaCl in buffer [80]. This medium was continued (flow rate 45 ml/h) until the optical density at 280 nm neared baseline (continuous line). At this point (arrow) NaCl was increased to 0.24 M, following which both phosphorylase phosphatase activity (○) and reductase reactivation activity (△) were eluted. The left ordinate is in units of phosphorylase phosphatase per ml of eluate; the right ordinate, in milliunits of reductase reactivation activity per ml [66].

formation of a protein inhibitor of reductase has been obtained the simplest explanation of these data was that the reductase itself was being dephosphorylated. This conclusion was later corroborated by the direct demonstration of Beg and co-workers that the inactive reductase is phosphorylated (see below) [87].

Studies by Philipp and Shapiro [88] and by Brown et al. [89] have provided additional support for this concept. Philipp and Shapiro [88] found that inactive reductase could be reactivated by potato acid phosphatase. At suboptimal doses, reactivation of reductase by the acid phosphatase and the reductase activator enzyme [77] was additive, while at higher doses neither enzyme was enhanced by the other. The two enzymes thus appear to have a common mechanism of action (i.e., dephosphorylation of the reductase). Brown et al. [89] have reported that inactive reductase can also be activated by alkaline phosphatase from *Escherichia coli*. It should be noted that quite large doses of the alkaline and acid phosphatase are required to activate the

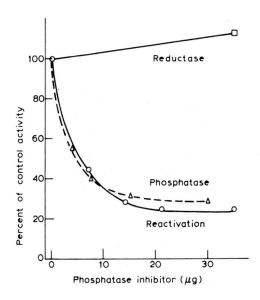

Figure 3. Parallel inhibition by the heat-stable phosphorylase phosphatase inhibitor protein of reductase reactivation and phosphorylase inactivation in the presence of added phosphorylase phosphatase. The inhibitor was purified from rat liver by the method of Khandelwal and Zinman [86]. Phosphorylase phosphatase was purified through the ethanol step as described by Ingebritsen et al. [66]. Phosphorylase phosphatase and reductase phosphatase assays were carried out as described by Ingebritsen et al.[66], except that all phosphatase assays were performed at 37 °C in a buffer composed of 5 mM DTT, 5 mM EDTA, 5 mM theophylline and 50 mM imidazole, pH 7.4

reductase, compared to the 35 000 dalton liver phosphoprotein phosphatase.

Reductase reactivation is severely inhibited by the two substrates of the reductase, NADPH and HMG-CoA. Nordstrom [90] observed that reactivation of inactive reductase catalyzed by the activating system from rat liver cytosol was completely blocked by 0.5 mM D,L-HMG-CoA. Saucier and Kandutsch [91] found that the activation of HMG-CoA reductase by an endogenous activator in L-cells and in fetal brain was severely inhibited by HMG-CoA. Half maximal inhibition was produced by D,L-HMG-CoA at a concentration of 5–10 μM. Free CoA was also inhibitory, but to a lesser extent. Hydroxymethylglutaric acid at levels as high as 80 μM was not inhibitory. The concentration of HMG-CoA producing half-maximal inhibition of activation is very similar to the K_m for HMG-CoA in the reductase catalyzed formation of mevalonate [4]. Thus, HMG-CoA may inhibit the reactivation as the result of a conformational change induced by binding of HMG-CoA to the active site of the enzyme. The concentration of HMG-CoA in whole rat liver is in the range of 20 μM [92]. Depending on the partitioning of HMG-CoA between cytosol and mitochondria, HMG-CoA may conceivably exert a redulatory influence on recuctase dephosphorylation and activation.

NADPH at a concentration of 20 mM was also a potent inhibitor of the activation [91]. At a lower concentration (2 mM), there was no effect [90]. The inhibitory effect of NADPH is probably not associated with its binding to the active site of the reductase as the K_m for this nucleotide (26 μM) is much lower than the inhibitory concentration. Since the total concentration of NADPH in rat liver is less than 1 mM [92], this nucleotide probably plays no role in the regulation of reductase activation.

5. Direct evidence for phosphorylation of reductase

Chow et al. [93] were the first to attempt a direct demonstration of reductase enzyme phosphorylation utilizing [γ-^{32}P]ATP(Mg) or adenylation with [*adenine*-^{14}C]ATP. The inactivated enzyme was precipitated with antibody to the most highly purified reductase available at the time. No label was recovered in the immunoprecipitate from either labelled substrate. The failure to demonstrate phosphorylation of the reductase in this early study may have resulted in part from the use of [γ-^{32}P]ATP with too low a specific radioactivity.

Bove and Hegardt [94] employing rat liver microsomes demonstrated that the rate of incorporation of ^{32}P from [γ-^{32}P]ATP into TCA-precipitable microsomal proteins was closely linked in time with the inactivation of reductase. Addition of the liver cytosol coordinately restored reductase activity and apparently removed covalently-bound ^{32}P. These studies, while encouraging, were complicated by the probable phosphorylation and dephosphorylation of many proteins in the membrane preparation.

Beg et al. [87] were able to recover inactivated, soluble reductase from rat liver microsomes labelled with ^{32}P from [γ-^{32}P]ATP. The rat liver enzyme was harvested with goat antibodies to homogeneous chicken liver reductase. A graded effect on inactivation and ^{32}P incorporation into reductase was obtained by using two concentrations of ATP. The label was stated to co-migrate with the protomer of reductase when resolved on SDS polyacrylamide gel. The action of phosphoprotein phosphatase was not reported in this communication [87]. A collaborative investigation between the laboratories of Rodwell and Rudney [95] has also demonstrated that reductase itself is phosphorylated under conditions in which the enzyme is inactivated by

$[\gamma\text{-}^{32}\text{P}]\text{ATP}$. In these studies, reductase was first inactivated on microsomes, then solubilized and purified to homogeneity. Reductase activity and ^{32}P were found to co-migrate with a single protein band on gel electrophoresis.

With the evidence at hand, it now seems appropriate to assign the name reductase kinase to the inactivating enzyme. The cycle of reactions depicted in Figure 4 is typical of a closed interconvertible system in which the state of phosphorylation of reductase, and thus its catalytic potential, would depend on the ratio of activities of reductase kinase and phosphoprotein phosphatase.

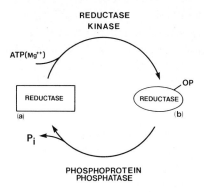

Figure 4. Monocyclic model for regulation of HMG-CoA reductase by reversible phosphorylation.

6. Properties of reductase kinase

The observation of Beg et al. [59] that microsomes could be rendered insensitive to ATP(Mg) inactivation of membrane-bound reductase by a relatively gentle extraction procedure pointed the way to a direct assay of reductase kinase. As an alternative procedure, Brown et al. [76] as well as Nordstrom et al. [77] employed microsomes that were pretreated by incubation at 37°C for several hours, since this apparently inactivated endogenous modulating enzyme. In either assay system fractions derived from microsomal extracts or the cytosol could be examined for reductase kinase activity. Beg et al. [59] and Brown et al. [76] found that soluble reductase kinase was precipitated with low concentrations of ammonium sulfate (33% saturation) and that it was non-dialyzable and heat labile [76]. Nordstrom et al. [77] partially purified a fraction from the cytosol by ammonium sulfate precipitation (35% saturation) and DEAE-sephadex chromatography. By contrast to the 35 000 dalton activating enzyme this protein exceeded 200 000 daltons. Later work indicated that the molecular weight lay between 360 000–380 000 [96,97]

Brown et al. [76] using reductase kinase from fibroblast extracts and Nordstrom et al. [77] using liver microsomal reductase kinase examined the nucleotide requirements for reductase inactivation. Both groups compared the relative potency of the nucleotide mono, di, and triphosphate derivatives of adenine, guanosine, inosine, uridine, and cytidine. Of the triphosphates, only ATP and GTP produced any inactivation of the reductase, although GTP was much less potent than ATP. Interestingly, ADP was almost as effective as ATP in producing inactivation of the reductase. The action of ADP may be explained by its rapid conversion to ATP by adenylate kinase endogenous to the microsomal preparations [76,78]. Supporting this idea is the observation by Brown et al. [76] that inclusion of an ADP regenerating system (glucose and hexokinase) effectively blocked the inactivation of reductase by ADP alone. Inactivation of soluble reductase with ADP was also diminished by AMP [98].

Several studies indicated that ADP (or a derivative nucleotide) was required beyond the requirement for ATP in reductase inactivation. Brown et al. [76] found that addition of either of two ATP regenerating systems (creatine kinase plus creatine phosphate or pyruvate kinase plus phosphoenolpyruvate) to the incubation system containing ATP prevented inactivation of the reductase. Nordstrom et al. [77] observed that optimal inactivation of soluble HMG-CoA reductase required the presence of an equimolar ratio of ATP:ADP. Since the synthesis of a nucleotide inhibitor of reductase from ADP was ruled out (see before), the most reasonable interpretation of these data is that ADP (or a derivative nucleotide) may influence the activity of either reductase kinase or reductase phosphatase. Relatively crude enzyme preparations were used in these studies and no attempt was made to inhibit endogenous phosphatase activity. Liver phosphorylase phosphatase is known to be inhibited by ATP, ADP, pyrophosphate and orthophosphate [99–106]. Conceivably inactivation of reductase would be facilitated by these phosphatase inhibitors. Alternatively, ADP (or a derivative nucleotide) may serve as a positive effector of reductase kinase.

Beg et al. [65] stated that solubilized reductase was capable of inactivation. Pursuing this in greater detail Nordstrom et al. [77] showed that reductase kinase would diminish the activity of reductase solubilized from microsomes and purified some 1400-fold. More recently, Ness [98] and Beg et al. [97] have shown that HMG-CoA reductase purified to homogeneity from rat liver is inactivated after incubation with MgATP and partially purified reductase kinase.

Ingebritsen et al. [78,96] have partially purified reductase kinase both from microsomal extracts and from liver cytosol, employing an assay system with microsomes depleted of inactivating activity by combining the procedures of Beg et al. [59] and Brown et al. [76]. Inactivation of the reductase in the deficient microsomes in the presence of MgATP was linear with time and with the amount of added reductase kinase fractions up to about 50% inactivation of the reductase. The K_m for ATP in the presence of 10 mM Mg^{2+} was 0.2 mM [96]. When the ATP concentration was fixed at 2 mM, maximal rates of inactivation were obtained at concentrations of free Mg^{2+} greater than 2 mM [78].

Beg and co-workers [97] have recently purified reductase kinase from microsomal extracts to apparent homogeneity. The enzyme possesses a molecular weight of 380 000. A single band of 58 000 daltons is observed on SDS gel electrophoresis.

Early studies [59,77] indicated that reductase kinase was present both in native microsomes and in the cytosol. Ingebritsen [78] has recently examined the subcellular distribution of the enzyme in more detail by comparing its distribution with that of cAMP-dependent protein kinase. The latter enzyme previously was shown to reside in the cytosol of rat liver [107]. These studies indicated that reductase kinase is also a cytosolic enzyme and that reductase kinase activity found in native microsomes is the result of adventitious binding of cytosolic protein to the membrane fraction. This conclusion is further supported by the observation that reductase kinase activity both in cytosol and in soluble, microsomal extracts exhibits very similar chromatographic behavior on DEAE-cellulose [108,109].

Ingebritsen et al. [78,96] also examined the question of whether reductase kinase was cAMP dependent. The effects of cAMP and the specific, heat-stable protein kinase inhibitor on reductase kinase and histone kinase activities were compared in both cytosolic and soluble microsomal reductase kinase preparations. In assays of both enzymes, conducted under identical incubation conditions, reductase kinase activity was unaffected by either cAMP or the protein kinase inhibitor. Nevertheless, the expected stimulation of histone kinase activity by cAMP and the inhibition by the kinase inhibitor were observed. Reductase kinase activity was not influenced by the addition of cGMP, cCMP or cIMP at concentrations of 10^{-5} M to 10^{-3} M [96]. Reductase kinase thus appeared to belong to a class of cyclic nucleotide independent protein kinases.

7. Distribution of modulating enzymes among animal tissues

The reductase modulating system has been shown to be present in a variety of animal species and in a number of cultured cell types. Hunter and Rodwell [110] have now completed an extensive study of the distribution of these enzymes among vertebrate livers. Both activating and inactivating systems were found in nine species (rat, mouse, gerbil, hamster, rabbit, pig, chicken, frog and catfish).

Brown et al. [76] demonstrated the presence of the reductase inactivating system in cultured human fibroblasts. This system was capable of inactivating not only HMG-CoA reductase from fibroblasts but also reductase present on rat liver microsomes. The properties of this fibroblast system (nucleotide specificity and fractionation with ammonium sulfate) are very similar to that of rat liver. Saucier and Kandutsch [91] have demonstrated the presence of both reductase activating and inactivating systems in L-cell cultures, in long-term cultures of fetal mouse liver cells, as well as in fetal mouse brains.

Since the reductase modulating enzymes are present both in the liver and in extrahepatic tissues of a variety of animal species, this system appears to have been well conserved in vertebrate species from an evolutionary standpoint. Hunter and Rodwell [110] noted that modulation of reductase activity, in quantitative terms, was a more constant feature than reductase activity itself. That is, the specific enzyme activity of fully activated reductase varied over a 50-fold range, whereas the response of ATP(Mg)-inactivated reductase to the addition of the reactivating enzyme (percent rise) varied only 3-fold. This circumstance of evolutionary conservation supports the view that reductase modulation through these enzymes may play a significant physiological role.

8. Modulation of reductase kinase

The scheme presented in Figure 4 must be changed to accommodate new evidence recently reported by Ingebritsen et al. [66], viz. the kinase enzyme itself is converted reversibly from an active to a less active catalytic state.

This concept was derivative from the observation that microsomes may become insensitive to ATP(Mg) by preincubation at 37 °C for several hours [76,77]. Ingebritsen and co-workers [66] observed that this time-dependent inactivation of microsomal reductase kinase was totally blocked by adding 50 mM NaF to the incubation medium. Control studies showed that fluoride had no effect on either reductase or reductase kinase activity per se. Moreover, it was found that the rate of inactivation of the microsomal enzyme was enhanced in phosphate-free buffer. Addition of phosphate back to the latter buffer slowed the rate of inactivation. Since fluoride and phosphate are both known inhibitors of phosphoprotein phosphatase these data raised the possibility that reductase kinase was also an interconvertible enzyme the active form of which was phosphorylated. Inactivation of the reductase kinase thus resulted from the presence of a phosphoprotein phosphatase in the microsomal preparation.

This notion was confirmed with the observation that purified phosphorylase phosphatase from rat liver markedly enhanced the rate of inactivation of reductase kinase present in microsomal extracts (Table 2). Both the slow basal inactivation of reductase kinase in this preparation and the phosphatase-stimulated inactivation were completely blocked by 50 mM NaF [66]. Similarly it was found that reductase kinase present in the cytosol was also subject to inactivation by the action of phosphatase. In the case of the cytosol, the basal rate of inactivation due to endogenous phosphatase was much higher than in microsomal extracts. (Compare Table 2 and Figure 5.)

TABLE 2

Inactivation of reductase kinase by phosphorylase phosphatase

Halide	Phosphatase (nanograms)	Time (min)	RK Activity (units/mg)	Percent inactivation
F^-, Cl^-	0	0	0.47	0
F^-, Cl^-	42	0	0.42	10.6
F^-, Cl^-	84	0	0.43	8.5
F^-	0	30	0.47	0
F^-	42	30	0.44	6.4
F^-	84	30	0.44	6.4
Cl^-	0	30	0.39	17.0
Cl^-	42	30	0.23	51.1
Cl^-	84	30	0.15	68.1
F^-	0	120	0.47	0
Cl^-	0	120	0.32	31.9

A soluble reductase kinase (RK) preparation (46.4 μg) was preincubated at 37 °C for the indicated time with either NaF or NaCl (5 μmol) and with the indicated nanogram quantities of phosphorylase phosphatase in imidazole buffer (final volume 60 μl). RK activity was then assayed immediately after bringing the concentrations of both NaF and NaCl in all samples to 50 mM. The phosphorylase phosphatase (36.6 units/mg) was purified on DEAE-Sephadex. Percent inactivation is relative to an RK activity of 0.47 units/mg [66].

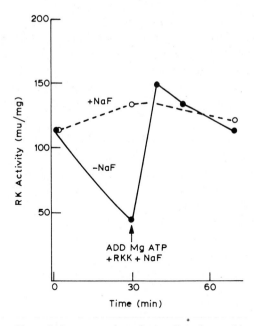

Figure 5. Interconversion of cytosolic reductase kinase (RK) between active and inactive forms. Rat liver cytosol was incubated at 37 °C in the absence of NaF (●) to inactivate reductase kinase. After 30 min 50 mM NaF was added to block further inactivation, and the system was supplemented with 4 mM MgCl$_2$, 2 mM ATP and microsomal reductase kinase kinase (see text). Incubation was then continued to 70 min. As a control, cytosol was incubated in the presence of 50 mM NaF for 30 min (○). The system was then supplemented with 4 mM MgCl$_2$, 2 mM ATP and reductase kinase kinase. The incubation was continued to 70 min. At 0, 30, 40, 50 and 70 min appropriate aliquots of the incubation mixtures were assayed for reductase kinase activity as described by Ingebritsen, et al. [66].

Both the basal and phosphatase-stimulated inactivation of reductase kinase were blocked by NaF ([108,109]; Figure 5).

Several lines of evidence indicated that the inactivation of reductase kinase is catalyzed by phosphorylase phosphatase and not some other enzyme contaminating the preparation. Firstly, both activities were found to copurify through a 1500-fold purification of the phosphatase. The ratio of the two activities in the preparations was constant throughout the purification. Secondly, NaF was found to produce a parallel inhibition of both activities, half maximal inhibition being in the concentration range of 3–4 mM. Finally, both activities were severely depressed in the presence of either of the two heat-stable phosphorylase phosphatase inhibitors from skeletal muscle [78]. Since these inhibitors are highly specific for phosphorylase phosphatase [82] it is clear that the inactivation of reductase kinase was catalyzed by the same phosphatase.

These studies established that a protein dephosphorylation event accompanied the inactivation of reductase kinase. The simplest interpretation of these data was that reductase kinase was an interconvertible enzyme the active form of which was phosphorylated.

Recognizing that a meaningful control circuit would require reactivation of reductase a search was made for an ATP(Mg)-dependent protein kinase. This kind of activity was found in microsomal extracts (and in the liver cytosol). To demonstrate this conversion, microsomes were first preincubated at 37°C to inactivate reductase kinase. Then the soluble, inactive enzyme was extracted from the microsomes, fluoride was added to block endogenous protein phosphatase, and the system supplemented with ATP(Mg). Under these conditions reductase kinase was reactivated in a time-dependent manner. It was discovered that the relatively high ionic strengths ordinarily used in the reductase enzyme assays blocked the reactivation of reductase kinase [66]. Since the reductase kinase assay itself can be conducted at the higher ionic strength, the evidence suggested that reductase kinase reactivation is not auto-catalyzed by reductase kinase. The new kinase, for want of a better term, has been designated reductase kinase kinase. The inhibition of reductase kinase kinase but not reductase kinase in high ionic strength buffer provided a means of distinguishing the two enzymes in the reductase kinase assay, since the activity of both enzymes was eventually assessed through changes in reductase activity.

Like reductase kinase, the kinase kinase is concentrated in the cytosol, although easily detectable quantities are adsorbed on freshly isolated microsomes. In the cytosol, the two enzymes are separable on DEAE cellulose. However, only reductase kinase has been prepared free of the other enzyme [111]. In a previous report Gibson and Ingebritsen [1] noted that reductase kinase kinase activity was present in the particulate glycogen protein complex obtained from rat liver extracts by precipitation at pH 6.1; resuspension in pH 7.4 buffer; and centrifugation at 100 000 × g. A more careful examination of the composition of this fraction [78] revealed that it also contained microsomes (using HMG-CoA reductase as a marker). Separation of these two components demonstrated that the reductase kinase kinase activity was associated with the microsomes rather than with the glycogen per se.

The reversibility of the reductase kinase interconversion is illustrated in Figure 5. Here cytosolic reductase kinase was inactivated by incubation with a phosphatase endogenous in the cytosolic preparation. After 30 min the inactivation was stopped by adding NaF and the incubation was continued in the presence of added reductase kinase kinase and MgATP. Complete reactivation of reductase kinase occurred within ten minutes. As a control, cytosolic reductase kinase which had been preincubated with NaF to prevent inactivation of the enzyme was further incubated with MgATP and reductase kinase kinase. No change in reductase kinase activity was noted. Thus, incubation with MgATP and reductase kinase kinase reverses the inactivation of the reductase kinase but does not alter the catalytic efficiency of the fully active enzyme. The interconversion of the reductase kinase is a fully reversible process.

These results led Ingebritsen et al. [1,66] to suggest that HMG-CoA reductase is modulated by the bicyclic system shown in Figure 6. In this system, the activities of both HMG-CoA reductase and reductase kinase are regulated by reversible phosphorylation. HMG-CoA reductase is active in the dephosphorylated state, whereas reductase kinase is inactive under these conditions. Dephosphorylation of both HMG-CoA reductase and reductase kinase is catalyzed by the 35 000 dalton liver phosphorylase phosphatase. Phosphorylation of reductase kinase requires MgATP and an enzyme, designated reductase kinase kinase, which is present in the cytosol and microsomes.

Recently, Beg and co-workers [97] have provided direct confirmation that

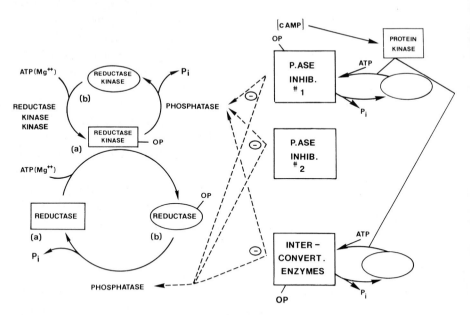

Figure 6. Bicyclic model for regulation of both HMG-CoA reductase and reductase kinase by reversible phosphorylation. Also shown are proposed routes for regulation of the bicyclic system by phosphatase inhibitors 1 and 2 by other phosphorylated proteins.

reductase kinase activity is modulated through reversible phosphorylation. After inactivation of homogeneous reductase kinase with phosphatase, the enzyme was incubated with cytosolic reductase kinase kinase, $[\gamma\text{-}^{32}P]ATP$ and Mg^{2+}. A time dependent increase in both reductase kinase activity and ^{32}P incorporation into the enzyme was observed. Incubation of ^{32}P-labelled reductase kinase with phosphatase resulted in loss of activity and the release of $[^{32}P]$orthophosphate into the incubation medium. After gel electrophoresis (non-denaturing conditions) of the ^{32}P-labelled reductase kinase, all of the radioactivity and the enzymatic activity were found to migrate with the protein band.

9. Regulation of the bicyclic system

As noted previously, glucagon and cAMP diminish reductase activity both in vivo and in isolated hepatocytes. Because of the analogy of the bicyclic reductase system with the classical system for activation of phosphorylase a, Ingebritsen [78] examined the possibility that reductase kinase kinase was regulated by cAMP. As noted before, reductase kinase is cAMP-independent. Using a similar approach, it was found that neither cAMP nor the heat stable inhibitor protein of cAMP-dependent protein kinase produced any effects on kinase kinase activity in the cytosol, microsomes or microsomal extracts. These modulators did however have the expected effects on cAMP-dependent protein kinase activity endogenous in these fractions. The possible relationship between these kinases and other cyclic nucleotide-independent protein kinases remains to be established.

The bicyclic reductase system is ideally suited for regulation through the 35 000 dalton liver phosphorylase phosphatase. This phosphatase catalyzes the dephosphorylation of both HMG-CoA reductase and reductase kinase (Figure 6). Inhibition of the phosphatase would produce a coordinate response in which the activity of reductase kinase would be increased and the activity of reductase would be decreased. Conversely, stimulation of the phosphatase would inhibit reductase kinase and increase reductase activity. As discussed by Stadtman and Chock [112], a bicyclic system would amplify a signal mediated through phosphatase effectors.

The dephosphorylation of reductase and reductase kinase by liver phosphorylase phosphatase is in keeping with the fact that this phosphatase can accept a variety of phosphorylated substrates including glycogen synthase b, phosphorylase kinase a [79,81,113,117], and adipose cell hormone-sensitive lipase [114]. In addition, a similar enzyme from skeletal muscle, referred to as phosphoprotein phosphatase 1 [115], has been shown to dephosphorylate acetyl-CoA carboxylase from mammary gland [116] in addition to phosphorylase *a*, glycogen synthase *b* and phosphorylase kinase *a* [115]. These observations have led to the suggestion that a single phosphoprotein phosphatase, referred to by various groups as phosphoprotein phosphatase 1 [115] or phosphoprotein phosphatase C (117), catalyzes the dephosphorylation of a number of key enzymes in glycogen and lipid metabolism in a coordinate manner [63].

Phosphorylase phosphatase activity in fresh extracts of liver and skeletal muscle resides principally with a high molecular weight (125 000–240 000) species [79,115, 117–119]. On treatment with ethanol, trypsin, or urea a stable, ultra-active 35 000 dalton subunit emerges. The high molecular weight species is thought to arise either as a result of the interaction of the 35 000 dalton subunit with a regulatory subunit [79,117,120,121] or as a result of aggregation of the small subunits [115]. It has been suggested that the regulatory subunit may endow substrate specificity on the enzyme [119]. Recent work by Ingebritsen [78] has shown that the liver high molecular weight phosphatase has the same relative activity towards reductase and reductase kinase (with respect to phosphorylase phosphatase) as the low molecular weight species. Thus, at least with reductase, reductase kinase and phosphorylase as substrates, the presence of the putative regulatory subunit does not alter the substrate specificity of the phosphatase.

The concept, that a single phosphoprotein phosphatase catalyzes the coordinate dephosphorylation of key enzymes in glycogen and lipid metabolism, has been recently challenged by several groups who have described a specific glycogen synthase phosphatase present in liver and cardiac muscle which is distinct from phosphoprotein phosphatase C [122–126]. No evidence has been obtained, however, to suggest that separate reductase and reductase kinase phosphatases are present in rat liver. Rather recent studies by Gil et al. [127] support the notion that phosphoprotein phosphatase C is the major reductase phosphatase in this tissue. These workers observed that all of the reductase phosphatase activity in rat liver cytosol was associated with two peaks of activity separated on DEAE-cellulose. Each of these phosphatase species was found to have a molecular weight of about 200 000. Treatment with room temperature ethanol converted each species to a low molecular weight form in the range of 30 000 daltons. Both peaks obtained from the DEAE-cellulose column were also active using phosphorylase *a* as a substrate. These properties are very similar to those previously reported for the native, high molecular weight form of phosphoprotein phosphatase C in rat liver cytosol [117,119].

Although the kinases in the bicyclic system are cAMP-independent, an alternate route for regulation of the system by glucagon and cAMP is feasible. Huang and Glinsmann [83], and more recently Nimmo and Cohen [82,128] have described a heat-stable protein of relatively low molecular weight in skeletal muscle which acts as a potent, specific inhibitor of the phosphoprotein phophatase-1 or -C (i.e., the 35 000 dalton phosphorylase phosphatase). This protein, identified as phosphatase inhibitor-1, is itself regulated by reversible phosphorylation. The protein is inhibitory only in the phosphorylated state. Phosphorylation of the inhibitor is catalyzed by cAMP-dependent protein kinase. Interestingly, the dephosphorylation of this inhibitor is catalyzed by phosphoprotein phosphatase-1 [82]. Dephosphorylation of inhibitor-1 appears to be the only dephosphorylation reaction of phosphoprotein phosphatase-1 not inhibited by this inhibitor protein. Recently Tao et al. [129] and Faulkes and Cohen [130] have shown that the activity of the inhibitor in rat hind limb [129] and the state of phosphorylation of the inhibitor in rabbit muscle [130] is markedly enhanced after treatment of the animals with epinephrine.

This inhibitor has been identified in skeletal muscle preparations from rat [83] and rabbit [82,128] and in liver from dogs [85]. Although the presence of the inhibitor in rat liver has not yet been demonstrated, Ingebritsen et al. (authors' unpublished observations) have recently obtained evidence for its existence in rabbit liver.

If this inhibitor is present in rat liver, then a highly sensitive mechanism (see Figure 6) exists for hormonal regulation of HMG-CoA reductase through cAMP as a second messenger. In this scheme, cAMP would act by stimulating the phosphorylation (and activation) of phosphatase inhibitor-1 by cAMP-dependent protein kinase. Since this protein inhibits the dephosphorylation of both HMG-CoA reductase and reductase kinase by phosphorylase phosphatase (see before), the activation of the inhibitor would lead to an increase in reductase kinase activity which, coupled with the inhibition of reductase dephosphorylation, would effect the inactivation of reductase.

One further route may exist for regulation of the system by glucagon and cAMP (Figure 6). If the hypothesis is correct that phosphorylase phosphatase acts in vivo as a general phosphoprotein phosphatase, then various phosphorylated protein substrates would be expected to compete for the available phosphatase activity. If the level of one or more of the phosphorylated substrates were increased, then the activity of the phosphatase against other substrates might well be diminished. Since cAMP-dependent protein kinase catalyzes the phosphorylation either directly or indirectly of phosphorylase b, phosphorylase kinase and glycogen synthase, an increase in cAMP levels in the cell would be expected to increase the concentration of the phosphorylated forms of these enzymes. The increased competition for available phosphatase which would result could lead to an inactivation of HMG-CoA reductase in a manner similar to that described for interaction of phosphatase inhibitor-1 with the system.

Insulin may also play a role in regulating the bicyclic system. As early as 1970, Bishop [131] and Gold [132] reported that liver phosphoprotein phosphatase activity was diminished in diabetic rats. Furthermore, treatment of these rats with insulin, restored phosphatase activity. These observations have been confirmed and extended in a number of other laboratories [133–136]. Although most of these studies have been performed using glycogen synthase b as substrate, Khandelwal et al. [134] have obtained similar results using phosphorylase a as a substrate. Recently, Miller [137] has reported similar effects of diabetes and insulin on phosphoprotein phosphatase in cardiac tissue.

In diabetic animals the recovery of phosphatase activity both in heart and liver is relatively slow, following the administration of insulin, requiring several hours for full restoration [131,132,137]. Moreover, the increase in phosphatase activity is prevented by treatment of the animals with cycloheximide [135–137] suggesting that insulin induces the synthesis of phosphatase. It has not been established, however, whether the newly synthesized protein is the phosphatase itself or a phosphatase modulator protein.

Khandelwal and co-workers [134] have found that the level of total phosphatase inhibitor activity in rat liver was increased in diabetic rats and that insulin administration restored inhibitor activity to normal levels. Interestingly, the effects of diabetes and insulin on phosphoprotein phosphatase activity were most pronounced when the activity was measured using endogenous substrates in concentrated liver extracts. When extracts were diluted and the activity assessed using added substrates, the effects of insulin were less impressive. These results are consistent with a role for the inhibitors in the regulation of liver phosphatase activity. If diminution of liver glycogen synthase and phosphorylase phosphatase activities in the diabetic state are indicative of similar changes in the activity of phosphoprotein phosphatase toward HMG-CoA reductase and reductase kinase, then the large decrease in reductase activity seen in diabetic rats [48] may at least in part be due to a decrease in the activation state of the reductase (Figure 6).

Similarly the increase in reductase activity seen upon insulin administration to normal or diabetic rats [47,51,52] may result from an activation of the enzyme (i.e., dephosphorylation) due to an enhancement of phosphatase activity.

10. State of activity of reductase in vivo

The interconvertible enzymes that regulate glycogen levels in muscle were the first to be characterized in terms of the relationship between their state of phosphorylation in vivo and expressed catalytic activity. Investigators as early as the decade of 1960 were well aware of the necessity for stopping the action of phosphoprotein phosphatases during the course of preparing tissue extracts for analysis. This was achieved by adding fluoride to the homogenizing medium. EDTA was also introduced to chelate the Mg^{2+} necessary for protein kinase action [138–140].

Nordstrom et al. [77] applied this approach to assessing the activation state of HMG-CoA reductase in vivo. Microsomes were isolated from liver homogenates prepared in media containing fluoride. Microsomal reductase was then assayed before and after complete activation of the enzyme with phosphatase. Surprisingly, it was found that only 15% of the full activity was apparently expressed in vivo. When microsomes were isolated without fluoride in the medium, it was found that more than 70% of full activity was expressed. Inclusion of EDTA along with fluoride did not alter the fraction of full activity expressed compared with fluoride alone. Significantly, after incubation of reductase (isolated with or without fluoride) with phosphatase, the same maximal reductase activity was expressed. These results suggested that HMG-CoA reductase was present in vivo largely in an inactivated state and that during the isolation of microsomes as ordinarily performed, i.e., without fluoride, the enzyme was progressively reactivated presumably due to the action of cellular phosphatase.

Using this approach, Ingebritsen and co-workers [141] have studied the effects of insulin and glucagon on the activation state of HMG-CoA reductase in rat hepatocytes. These cells were isolated from rats at the time of the normal diurnal maximum of reductase. As in whole liver, reductase in freshly isolated liver cells was present largely in the inactive form. However, during the course of a 2.5-h incubation of the cells in suspension, the activation state rose from 20% initially to as much as 100% by the end of the incubation (basal activation). This activation of the enzyme which required less than 1 h, was preceded by a lag period of variable length in which little activation was observed. When 10 nM glucagon was added to the incubation, the basal activation was blocked. Addition of 85 nM insulin in place of glucagon markedly enhanced the rate of activation.

In the same studies, expressed reductase kinase activity was assayed in hepatocyte cytosol (isolated in the presence of fluoride and EDTA). Reductase kinase activity varied inversely to expressed reductase activity in response to insulin and glucagon. Insulin diminished whereas glucagon enhanced the kinase activity.

These results provide strong support for the concept that insulin and glucagon regulate hepatic reductase through the bicyclic modulating system. As previously discussed, the most likely route for this regulation is through the phosphoprotein phosphatase which dephosphorylates both reductase and reductase kinase. The possibility that insulin has a more direct effect on the two kinases in the system has not been excluded, however.

The basal activation of reductase which was observed during incubation of the hepatocytes may result from the removal of normal hormonal signals which impinge upon the liver in the intact animal. Lin and Snodgrass [142] observed a similar activation of reductase in hepatocytes maintained for several days in culture in the absence of hormones. In addition, Brown et al. [89] stated that reductase in cultured fibroblasts was present in the fully active form.

In contrast to the acute response of isolated hepatocytes to insulin and glucagon, Brown et al. [89] observed little change in the low state of activation of liver reductase in rats subjected to a number of long term manipulations. The rates of cholesterol synthesis in these studies were varied by subjecting the animals to diurnal cycling, feeding, fasting, cholesterol feeding, cholestyramine feeding, and stress. Although the rate of cholesterol synthesis and the maximal reductase activity was strongly correlated in a linear manner over a 50-fold range in rates, the fraction of the enzyme in the active state was in all cases between 7–16%. A direct comparison between these and the hepatocyte studies may not be warranted, however, since the state of reductase in hepatocytes can be abruptly fixed in fluoride and EDTA without interference from possible neuroendocrine disturbances attending the handling and sacrifice of whole animals.

As discussed at the outset of this review, feedback regulation of cholesterol biosynthesis and HMG-CoA reductase by cholesterol has been the subject of investigation in a number of laboratories. It is clear from these studies that long term regulation of reductase occurs through induction and repression of enzyme synthesis. Recently, however, several groups have suggested that cholesterol may also exert a short term regulatory influence on reductase by mechanisms other than regulation of reductase synthesis. Higgins and Rudney in 1973 [26] compared the effects of cholesterol feeding on rat liver HMG-CoA reductase activity and on the yield of solubilized reductase precipitated by antibody prepared against the most highly purified preparation of the enzyme then available. Although this study may be open to criticism since the antibody was not prepared against homogeneous enzyme, it was found that reductase activity declined much more rapidly than did the amount of reductase protein after cholesterol feeding.

Edwards et al. [34] observed that the injection of mevalonolactone into rats produced a fall in hepatic reductase that was much too rapid (85% decrease in 45 min) to be accounted for by inhibition of reductase synthesis (half-life = 2–4 h). Here, administration of mevalonolactone increased cholesterol biosynthesis by providing substrate that by-passed reductase. These two studies showed that cholesterol may exert a short term modulatory influence on the reductase activity.

That at least part of this short term regulation may be accounted for by changes in the activation state of reductase is suggested by several recent investigations. Arebalo et al. [35] isolated soluble reductase from the liver of control rats and of rats subjected to short term cholesterol feeding (6 h). After partial purification (estimated at 3% of homogeneity) the enzyme from the cholesterol-fed rats had about 1/5 of the specific activity observed in controls. Furthermore, after treatment with phosphoprotein phosphatase, the reductase activity from control rats was increased 2-fold, whereas that from the cholesterol-fed rats was increased 4-fold.

Cholestyramine is known to produce a large enhancement (10-fold) of microsomal reductase activity [143] by promoting the efflux of cholesterol from the liver after conversion to bile acids [144]. It has not been established whether these changes reflect a net increase in reductase protein, an increase in the catalytic efficiency of the enzyme or both. Edwards et al. [17] and Srikantaiah et al. [36] have observed that reductase purified from cholesty-ramine-fed animals consistently has a 2-fold higher specific activity than that from control animals. The difference between the crude and purified preparations may be due to activation of the enzyme by contaminating phosphatases during its isolation.

Although these studies suggest that cholesterol may regulate the activation state of reductase, further work is needed to establish the point. In the studies cited above, no attempt was made to fix the activation state of the enzyme using fluoride and EDTA. In addition, studies need to be done to determine whether cholesterol affects the activities of reductase kinase, reductase kinase kinase, reductase phosphatase or reductase kinase phosphatase.

Acknowledgments

The authors wish to acknowledge the skillfull technical assistance of R.A. Parker in the research cited from their laboratory. These studies were

90

supported by grants from the National Institutes of Health (AM 19299 and AM 21278); American Heart Association, Indiana Affiliate; and the Grace M. Showalter Foundation.

References

1. Gibson, D.M. and Ingebritsen, T.S. (1978) Life Sci., 23, 2649–2664.
2. Brown, W.E. and Rodwell, V.W. (1979) in: Dehydrogenases Requiring Nicotinamide Coenzymes (Jeffery, J.J., ed.), pp. 232–272, Birkhauser-Verlag, Basel, Switzerland.
3. Dugan, R.E. and Porter, J.W. (1977) in: Biochemical Actions of Hormones (Litwack, G., ed.), Vol. 4, pp. 197–247, Academic Press, New York.
4. Rodwell, V.W., Nordstrom, J.L. and Mitschelen, J.J. (1976) Adv. Lipid Res., 14, 1–74.
5. Beytia, E.D. and Porter, J.W. (1976) Ann. Rev. Biochem., 45, 113–142.
6. Dempsey, M.E. (1974) Ann. Rev. Biochem., 43, 967–990.
7. Bortz, W.M. (1973) Metab. Clin. Exp., 22, 1507–1524.
8. Holloway, P.W. (1970) in: Lipid Metabolism (Wakil, S.J., ed.), pp. 371–429, Academic Press, New York.
9. Goldfarb, S. (1972) FEBS Lett., 24, 153–155.
10. Heller, R.A. and Gould, R.G. (1973) Biochem. Biophys. Res. Commun., 50, 859–864.
11. Brown, M.S., Dana, S.E., Dietschy, J.M. and Siperstein, M.D. (1973) J. Biol. Chem., 248, 4731–4738.
12. Singer, S.J. and Nicolson, G.L. (1972) Science, 175, 720–731.
13. Heller, R.A. and Gould, R.G. (1974) J. Biol. Chem., 249, 5254–5260.
14. Tormanen, C.D., Srikantaiah, M.V., Hardgrave, J.E. and Scallen, T.J. (1977) J. Biol. Chem., 252, 1561–1565.
15. Racker, E. (1967) Fed. Proc., 26, 1335–1340.
16. Kleinsek, D.A., Ranganathan, S. and Porter, J.W. (1977) Proc. Natl. Acad. Sci. USA, 74, 1431–1435.
17. Edwards, P.A., Lemongello, D. and Fogelman, A.M. (1978) Fed. Proc., 37, 1524.
18. Ness, G.C., Spindler, C.D. and Moffler, M.H. (1979) Fed. Proc., 38, 672.
19. Beg, Z.H., Stonik, J.A. and Brewer, H.B. (1977) FEBS Lett., 80, 123–129.
20. Brown, M.S. and Goldstein, J.L. (1974) J. Biol. Chem., 249, 7306–7314.
21. Chen, H.W., Kandutsch, A.A. and Waymouth, C. (1974) Nature (London), 251, 419–421.
22. Jackson, R.L., Morrisett, J.D. and Gotto, A.M. (1976) Physiol. Rev., 56, 259–316.
23. Myant, N.B. and Mitropoulos, K.A. (1977) J. Lipid Res., 18, 135–153.
24. Goldstein, J.L. and Brown, M.S. (1977) Ann. Rev. Biochem., 46, 897–930.
25. Andersen, J.M., Turley, S.D. and Dietschy, J.M. (1979) Proc. Natl. Acad. Sci. USA, 76, 165–169.
26. Higgins, M. and Rudney, H. (1973) Nature New Biol., 246, 60–61.
27. Beirne, O.R., Heller, R. and Watson, J.A. (1977) J. Biol. Chem., 252, 950–954.
28. Brown, M.S., Dana, S.E. and Goldstein, J.L. (1974) J. Biol. Chem., 249, 789–796.
29. Dugan, R.E., Slakey, L.L., Briedis, A.V. and Porter, J.W. (1972) Arch. Biochem. Biophys., 152, 21–27.
30. Edwards, P.A. and Gould, R.G. (1972) J. Biol. Chem., 247, 1520–1524.
31. Hickman, P.E., Horton, G.J. and Sabine, J.R. (1972) J. Lipid Res., 13, 17–22.
32. Higgins, M., Kawachi, T. and Rudney, H. (1971) Biochem. Biophys. Res. Commun., 45, 138–144.
33. Rodwell, V.W., McNamara, D.J. and Shapiro, D.J. (1973) Adv. Enzymol. Relat. Areas Mol. Biol., 38, 373–412.
34. Edwards, P.A., Popjak, G., Fogelman, A.M. and Edmond, J. (1977) J. Biol. Chem., 252, 1057–1063.

35. Arebalo, R.E., Hardgrave, J.E. and Scallen, T.J. (1979) Fed. Proc., 38, 632.
36. Srikantaiah, M.V., Tormanen, C.D., Redd, W.L., Hardgrave, J.E. and Scallen, T.J. (1977) J. Biol. Chem., 252, 6145–6150.
37. Kandutsch, A.A. and Saucier, S.E. (1969) J. Biol. Chem., 244, 2299–2305.
38. Berndt, J., Gaumert, R. and Lowel, M. (1972) Hoppe-Seyler's Z. Physiol. Chem., 353, 1454–1460.
39. Edwards, P.A. and Gould, R.G. (1972) J. Biol. Chem., 247, 1520–1524.
40. Edwards, P.A., Muroya, H. and Gould, R.G. (1972) J. Lipid Res., 13, 396–401.
41. Huber, J., Hamprecht, B., Muller, O.-A. and Guder, W. (1972) Hoppe-Seyler's Z. Physiol. Chem., 353, 313–317.
42. Booth, R., Gregory, K.W. and Smith, C.Z. (1972) Biochem. J., 130, 72p.
43. Edwards, P.A. and Gould, R.G. (1974) J. Biol. Chem., 249, 2891–2896.
44. Slakey, L.L., Craig, M.C., Beytia, E., Briedis, A., Feldbruegge, D.H., Dugan, R.E., Qureshi, A.A., Subbarayan, C. and Porter, J.W. (1972) J. Biol. Chem., 247, 3014–3022.
45. Shapiro, D.J. and Rodwell, V.W. (1969) Biochem. Biophys. Res. Commun., 37, 867–872.
46. Back, P., Hamprecht, B. and Lynen, F. (1969) Arch. Biochem., Biophys., 133, 11–21.
47. Lakshmanan, M.R., Nepokroeff, C.M., Ness, G.C., Dugan, R.E. and Porter, J.W. (1973) Biochem. Biophys. Res. Commun., 50, 704–710.
48. Nepokroeff, C.M., Lakshmanan, M.R., Ness, G.C., Dugan, R.E. and Porter, J.W. (1974) Arch. Biochem. Biophys., 160, 387–393.
49. Dugan, R.E., Ness, G.C., Lakshmanan, M.R., Nepokroeff, C.M. and Porter, J.W. (1974) Arch. Biochem. Biophys., 161, 499–504.
50. Laskhmanan, M.R., Dugan, R.E., Nepokroeff, C.M., Ness, G.C. and Porter, J.W. (1975) Arch. Biochem. Biophys., 168, 89–95.
51. Huber, J., Guder, W., Latzin, S. and Hamprecht, B. (1973) Hoppe-Seyler's Z. Physiol. Chem., 354, 795–798.
52. White, L.W. (1972) Circulation, 46, Suppl. 2, II–253.
53. Geelen, M.J.H. and Gibson, D.M. (1975) FEBS Lett., 58, 334–339.
54. Geelen, M.J.H. and Gibson, D.M. (1976) in: Use of Isolated Liver Cells and Kidney Tubules in Metabolic Studies, pp. 219–230, North-Holland, Amsterdam.
55. Bhathena, S.J., Avigan, J. and Schreiner, M.E. (1974) Proc. Natl. Acad. Sci. USA, 71, 2174–2178.
56. Ma, G.Y., Gove, C.D. and Hems, D.A. (1978) Biochem. J., 174, 761–768.
57. Edwards, P.A., Lemongello, D. and Fogelman, A.M. (1979) J. Lipid Res., 20, 2–7.
58. Bricker, L.A. and Levey, G.S. (1972) J. Biol. Chem., 247, 4914–4915.
59. Beg, Z.H., Allmann, D.W. and Gibson, D.M. (1973) Biochem. Biophys. Res. Commun., 54, 1362–1369.
60. Bloxham, D.P. and Akhtar, M. (1971) Biochem. J., 123, 275–278.
61. Misbin, R.I., Capuzzi, D.M. and Margolis, S. (1972) Horm. Metab. Res., 4, 176–178.
62. Edwards, P.A. (1975) Arch. Biochem. Biophys., 170, 188–203.
63. Killilea, S.D., Brandt, H. and Lee, E.Y.C. (1976) Trends Biochem. Sci., 1, 30–33.
64. Robison, G.A., Butcher, R.W. and Sutherland, E.W. (1971) cAMP. Academic Press.
65. Beg, Z.H., Allmann, D.W., Anderson, P.J., Pruden, E. and Gibson, D.M. (1974) Fed. Proc., 33, 1428.
66. Ingebritsen, T.S., Lee, H.-S., Parker, R.A. and Gibson, D.M. (1978) Biochem. Biophys. Res. Commun., 81, 1268–1277.
67. Goodwin, C.D. and Margolis, S. (1973) Fed. Proc. 32, 671.
68. Goodwin, C.D. and Margolis, S. (1973) J. Biol. Chem., 248, 7610–7613.
69. Goodwin, C.D. and Margolis, S. (1973) Circulation, 48, Suppl. 4, 244.
70. Goodwin, C.D. (1975) Fed. Proc., 34, 548.
71. Goodwin, C.D. and Margolis, S. (1978) J. Lipid Res., 19, 747–756.
72. Berndt, J. and Gaumert, R. (1974) Hoppe-Seyler's Z. Physiol. Chem., 355, 905–910.
73. Berndt, J., Hegardt, F.G., Bove, J., Gaumert, R. Still, J. and Cardo, M.-T. (1976) Hoppe-

Seyler's Z. Physiol. Chem., 357, 1277–1282.

74. Hizukuri, S. and Larner, J. (1964) Biochemistry, 3, 1783–1788.

75. Shimazu, T., Tokutake, S. and Usami, M. (1978) J. Biol. Chem., 253, 7376–7382.

76. Brown, M.S., Brunschede, G.Y. and Goldstein, J.L. (1975) J. Biol. Chem., 250, 2502–2509.

77. Nordstrom, J.L., Rodwell, V.W. and Mitschelen, J.J. (1977) J. Biol. Chem., 252, 8924–8934.

78. Ingebritsen, T.S. (1979) Ph.D.Thesis, Indiana University.

79. Lee, E.Y.C., Brandt, H., Capulong, Z.L. and Killilea, S.D. (1976) Adv. Enz. Reg., 14, 467–490.

80. Brandt, H., Capulong, Z.L. and Lee, E.Y.C. (1975) J. Biol. Chem., 250, 8038–8044.

81. Killilea, S.D., Brandt, H., Lee, E.Y.C. and Whelan, W.J. (1976) J. Biol. Chem., 251, 2363–2368.

82. Nimmo, G.A. and Cohen, P. (1978) Eur. J. Biochem., 87, 341–351, 353–365.

83. Huang, F.L. and Glinsmann, W.H. (1975) Proc. Natl. Acad. Sci. USA, 72, 3004–3008.

84. Goris, J., Defreyn, G. and Merlevede, W. (1977) Biochem. Soc. Trans., 5, 978–979.

85. Goris, J., Defreyn, G., Vandenheede, J.R. and Merlevede, W. (1978) Eur. J. Biochem., 91, 457–464.

86. Khandelwal, R.L. and Zinman, S.M. (1978) J. Biol. Chem., 253, 560–565.

87. Beg, Z.H. Stonik, J.A. and Brewer, H.B. (1978) Proc. Natl. Acad. Sci. USA, 75, 3678–3682.

88. Philipp, B.W. and Shapiro, D.J. (1979) Fed. Proc., 38, 481.

89. Brown, M.S., Goldstein, J.L. and Dietschy, J.M. (1979) J. Biol. Chem., 254, 5144–5149.

90. Nordstrom, J.L. (1976) Ph.D. Thesis, Purdue University.

91. Saucier, S.E. and Kandutsch, A.A. (1978) Biochim. Biophys. Acta, 572, 541–556.

92. Williamson, D.H. and Brosnan, J.T. (1974) in: Methos of Enzymatic Analysis (Bergmeyer, H.U., ed.), Vol. 4, pp. 2266–2302.

93. Chow, J.C., Higgins, M.J.P. and Rudney, H. (1975) Biochem. Biophys. Res. Commun., 63, 1077–1084.

94. Bove, J. and Hegardt, F.G. (1978) FEBS Lett., 90, 198–202.

95. Rogers, D.H., Keith, M.L., Rodwell, V.W. and Rudney, H. (1979) Abstract, J. Lipid Res., in press.

96. Ingebritsen, T.S., Parker, R.A. and Gibson, D.M. (1979) Miami Winter Symposium 16, in press.

97. Beg, Z.H., Stonik, J.A. and Brewer, H.B. (1979) Abstract, J. Lipid Res., in press.

98. Ness, G.C. (1979) Abstract, J. Lipid Res., 20, 1048.

99. Torres, H.N. and Chelala, C.A. (1970) Biochim. Biophys. Acta, 198, 495–503.

100. Chelala, C.A. and Torres, H.N. (1970) Biochim. Biophys. Acta, 198, 504–513.

101. Kato, K., Kobayashi, M. and Sato, S. (1975) J. Biochem., 77, 811–815.

102. Hsiao, K.-J., Sandberg, A.R. and Li, H.-C. (1978) J. Biol. Chem., 253, 6901–6907.

103. Khatra, B.S. and Soderling, T.R. (1978) Biochem. Biophys. Res. Commun., 85, 647–654.

104. Khandelwal, R.L. (1978) Arch. Biochem. Biophys., 191, 764–773.

105. Jacob, A. and Diem, S. (1979) Biochim. Biophys. Acta, 567, 174–183.

106. Burchell, A. and Cohen, P. (1978) Biochem. Soc. Trans., 6, 220–222.

107 Chen, L.-J. and Walsh, D.A. (1971) Biochemistry, 10, 3614–3621.

108. Ingebritsen, T.S., Parker, R.A. and Gibson, D.M. (1979) J. Supramol. Struct. 9, Suppl. 3, 18.

109. Ingebritsen, T.S., Parker, R.A. and Gibson, D.M. (1979) Fed. Proc., 38, 481.

110. Hunter, C.F. and Rodwell, V.W. (1979) J. Biol. Chem., in press.

111. Ingebritsen, T.S, Parker, R.A. and Gibson, D.M. (1979) Abstracts 11th International Congress of Biochemistry, p. 304, Abstract 04-5-S118.

112. Stadtman, E.R. and Chock. P.B. (1978) Curr. Top. Cell. Reg., 13, 53–95.

113. Killilea, S.D., Lee, E.Y.C., Brandt, H. and Whelan, W.J. (1976) in: Metabolic Interconversion of Enzymes (Shaltiel, S., ed.), pp. 103–114, Springer-Verlag, Berlin.
114. Severson, D.L., Khoo, J.C. and Steinberg, D. (1977) J. Biol. Chem., 252, 1484–1489.
115. Cohen, P. (1978) Curr. Top. Cell. Reg., 14, 117–196.
116. Burchell. A., Foulkes, J.G., Cohem P.T.W., Condon, G.D. and Cohen, P. (1978) FEBS Lett., 92, 68–72.
117. Lee, E.Y.C., Mellgren, R.L., Killilea, S.D. and Aylward, J.H. (1977) in: Regulatory Mechanisms of Carbohydrate Metabolism (Esmann, V., ed.), FEBS Symposium, Vol. 42, pp. 327–346, Pergamon Press, New York.
118. Mellgren, R.L., Aylward, J.H., Killilea, S.D. and Lee, E.Y.C. (1979) J. Biol. Chem., 254, 648–652.
119. Killilea, S.D., Mellgren, R.L., Aylward, J.H., Metieh, M.E. and Lee, E.Y.C. (1979) Arch. Biochem. Biophys., 193, 130–139.
120. Brandt, H., Lee, E.Y.C. and Killilea, S.D. (1975) Biochem. Biophys. Res. Commun., 63, 950–956.
121. Brandt, H., Killilea, S.D. and Lee, E.Y.C. (1974) Biochem. Biophys. Res. Commun., 61, 598–604.
122. Binstock, J.F. and Li, H.-C. (1979) Biochem. Biophys. Res. Commun., 87, 1226–1234.
123. Laloux, M., Stalmans, W. and Hers, H.-G. (1978) Eur. J. Biochem., 92, 15–24.
124. Tan, A.W.H. and Nuttall, F.Q. (1978) Biochim. Biophys. Acta, 522, 139–150.
125. Kikuchi, K., Tamura, S., Hiraga, R.A. and Taniki, S. (1977) Biochem. Biophys. Res. Commun., 75, 29–37.
126. Tan, A.W.H. and Nuttall, F.Q. (1976) Biochim. Biophys. Acta, 445, 118–130.
127. Gil, G., Sitges, M. and Hegardt, F.G. (1979) Abstract, J. Lipid Res., in press.
128. Cohen, P., Nimmo, G.A. and Antoniw, J.F. (1977) Biochem. J., 162, 435–444.
129. Tao, S.-H. Huang, F.L., Lynch, A. and Glinsmann, W.H. (1978) Biochem. J., 176, 347–350.
130. Faulkes, J.G. and Cohen, P. (1979) Eur. J. Biochem., 97, 251–256.
131. Bishop, J.S. (1970) Biochim. Biophys. Acta, 208, 208–218.
132. Gold, A.H. (1970) J. Biol. Chem., 245, 903–905.
133. Nichols, W.K. and Goldberg, N.D. (1972) Biochim. Biophys. Acta, 279, 245–259.
134. Khandelwal, R.L., Zinman, S.M. and Zebrowski, E.J. (1977) Biochem. J., 168, 541–548.
135. Haverstick, D.M., Dickemper, D. and Gold, A.H. (1979) Biochem. Biophys. Res. Commun., 87, 177–183.
136. Miller, T.B. (1979) Biochim. Biophys. Acta, 583, 36–46.
137. Miller, T.B. (1978) J. Biol. Chem., 253, 5389–5394.
138. Danforth, W.H., Helmreich, E. and Cori, C.F. (1962) Proc. Natl. Acad. Sci. USA, 48, 1191–1199.
139. Stull, J.T. and Mayer, S.E. (1971) J. Biol. Chem., 246, 5716–5723.
140. Shimazu, T. and Amakawa, A. (1975) Biochim. Biophys. Acta, 385, 242–256.
141. Ingebritsen, T.S., Geelen, M.J.H., Parker, R.A., Evenson, K.J. and Gibson, D.M. (1979) J. Biol. Chem., 254, 9986–9989.
142. Linn, R.C. and Snodgrass, P.J. (1979) Abstracts 11th International Congress of Biochemistry, p. 304, Abstract 04-5-S116.
143. Goldfarb, S. and Pitot, H.C. (1972) J. Lipid Res., 13, 797–801.
144. Huff, J.W., Gilfillan, J.L. and Hunt, V.M. (1963) Proc. Soc. Exp. Biol. Med., 114, 352–355.

CHAPTER 5

Protein phosphorylation and steroidogenesis

GEORGE S. BOYD AND A.M.S. GORBAN

1. Introduction

Steroid hormones are produced in the adrenals, gonads and placenta. In the process of steroidogenesis, cholesterol plays an important part as an obligatory precursor of the steroid hormones. The tissues involved in steroidogenesis contain mitochondria capable of effecting a transformation of cholesterol into the key intermediate in steroid hormone production, pregnenolone [1]. This reaction is termed the cholesterol side chain cleavage reaction, and it occurs in the inner cristae of the adrenal mitochondria. The cholesterol side chain cleavage reaction occurs as a result of sequential hydroxylation in the sterol side chain at C_{22}, followed by hydroxylation at C_{20} and finally a splitting of the cholesterol side chain between C_{20} and C_{22} to yield pregnenolone and isocaproic aldehyde [2].

In the adrenal cortex, pregnenolone produced in the mitochondria is subsequently modified in the endoplasmic reticulum by 17α-hydroxylation and/or 21-hydroxylation and by oxidation of the 3β-hydroxysteroid Δ-5 structure to a 3-keto Δ-4 structure. It is therefore possible within the endo-plasmic reticulum of the adrenal cortex to convert pregnenolone into deoxycortisol or deoxycorticosterone; the production of these steroids is in part dependent upon the species. Deoxycortisol and deoxycorticosterone may then be transported back into mitochondria for 11β-hydroxylation or in some cases for 11β-hydroxylation and 18-hydroxylation. The key steps in the transformation of cholesterol into adrenal steroids are shown in Figure 1. There are several comprehensive reviews on this topic [2–4].

It is now well established that the process of steroidogenesis in the adrenal cortex is markedly affected by ACTH [5–16] and also by angiotensin II [6,17]. In this review we are concerned with the role of ACTH as an activator of steroidogenesis in adrenal cortex.

Cohen (ed.) Recently discovered systems of enzyme regulation by reversible phosphorylation

96

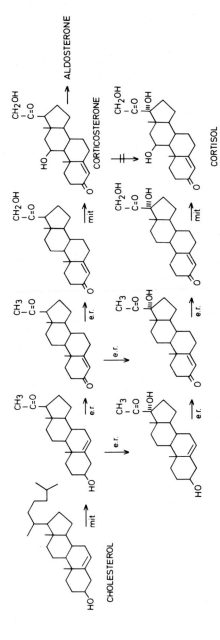

Figure 1. The key steps in the transformation of cholesterol into steroid hormones in the adrenal cortex.

When the plasma concentration of ACTH is raised, there is an increased output of adrenal steroids by the adrenal cortex. The problem which we wish to explore in this review concerns the links in the chain between the interaction of the trophic hormone with the plasma membrane of the cell in this steroidogenic tissue, and the ultimate increase in output of steroid hormones by the cell. Evidence is advanced to show that protein phosphorylation may play an important part in transducing the ACTH signal into an increment in steroid hormone output in the adrenal cortex.

2. The interaction of ACTH with the adrenal cells

The studies of Haynes and Berthet [8] showed that ACTH activated adrenal tissues with a resultant increase in the intracellular concentration of cyclic AMP. It was shown that this increase in cyclic AMP in adrenal tissues was causally related to the increased steroidogenesis by the tissue, because exogenous cyclic AMP could increase steroid hormone production in adrenal tissue in the absence of ACTH. These observations placed cyclic AMP firmly in the role of a secondary messenger in the interpretation of the ACTH signal and its transmission to the remainder of the steroid hormone production machinery in the cell.

The purified ACTH molecule consists of a single polypeptide chain of 39 amino acids [18,19]. It is now known that the complete 39 amino acid polypeptide chain is not required for activation of the adrenal cells with resultant production of steroid hormones. The work of Sayers and associates [12,20,21] and other [10,22,23] delineated the aspects of the ACTH molecule required to produce the adrenal cell steroidogenic response. These workers have shown that successful interaction with the plasma membrane receptor requires the first 24 amino acids of the ACTH molecule.

It has also been shown that there is no requirement for the polypeptide molecule to penetrate the cell membrane because, if the ACTH molecule is covalently attached to a high molecular weight synthetic polymer incapable of entering the adrenal cell, this modified, high molecular weight 'hormone' is able to elicit a satisfactory response from the adrenal tissue. These studies emphasise the existence of a specific receptor or receptors on the external surface of the plasma membrane of the adrenal cells capable of identifying specific aspects of the ACTH molecule [23–27]. The hormone–receptor event may involve activation of a membrane phospholipase, alterations to the membrane phospholipids and changes in the sequestration of membrane calcium ions [28]. This hormone–receptor interaction from the outside of the cell is transmitted to the inner aspect of the plasma membrane, resulting in activation of adenyl cyclase located in the inside of the plasma membrane [29–33].

When adrenal cells are exposed to ACTH either in vivo or in vitro the rate of change of concentration of cyclic AMP within the cells is very rapid [10,34–36]. Within 5 min there may be a 5-fold change in concentration of cyclic AMP in adrenal cells activated by ACTH.

By contrast, the change in concentration of cyclic GMP in these cells is negligible or very slight [16,37,38]. This observation has supported the generally accepted concept that it is cyclic AMP rather than cyclic GMP which is the key secondary messenger in adrenal cells.

It has been established that if adrenal cells or adrenal slices are exposed to cycloheximide or puromycin before being exposed to ACTH the response of these cells to ACTH, as judged by an increased production of cyclic AMP, is unaffected [10]. The effect of ACTH on the adrenal cell plasma membrane, resulting in activation of adenyl cyclase is therefore not dependent upon protein synthesis.

3. Receptors in the adrenal cell responding to cyclic AMP

There is now a vast body of evidence suggesting that cyclic AMP induces steroidogenesis in adrenal tissue, and much work has been directed towards attempting to discover how this cyclic nucleotide activates steroidogenesis [8,23,39]. In this connection, studies have been made on the possible receptors for cyclic AMP in the cell. It has been shown that cyclic AMP binds to a protein or proteins in the cytosol of the cell and to a protein in the endoplasmic reticulum of the adrenal cell [40,41].

The cytosolic cyclic AMP-binding protein has been studied in detail [40,42]. It has been shown that treatment of the binding fraction with trypsin or a protease destroys the ability of this fraction to bind cyclic AMP. On the other hand the treatment of this fraction with ribonuclease or deoxyribonuclease fails to affect the ability to bind cyclic AMP [40]. This evidence, in association with the observation that heating the fraction at 50 °C for 15 min destroyed the cyclic AMP binding properties, suggested that the binding material in the cell cytosol and in the microsomal fraction was proteinaceous in nature.

The bulk of the cyclic AMP-binding protein has been found in the cell cytosol. This suggests that the cyclic AMP generated in the vicinity of the plasma membrane of the cell is released into the cyctosol and bound or sequestered there by this cyclic AMP-binding protein. Evidence has been presented showing that the cyclic AMP-binding protein in the cell cytosol is identical to the protein found in the microsomes. It has been shown from equilibrium dialysis studies that the cyclic AMP-binding site on the protein is apparently a single non-interacting binding site with a binding constant of K_a = 5 × 10^{-8}.

The specificity of the binding protein for various nucleotides has been tested in the usual way by studying the ability of other nucleotides to compete with cyclic AMP for the active site of the binding protein. Various nucleotides such as 5'-AMP, 2'-AMP, 3'-AMP, cyclic 2',3'-AMP, ATP, 2'-GMP, 3'-GMP, GTP and cyclic 2',3'-CMP were unable to compete with cyclic AMP for the binding protein [40,42]. On the other hand, cyclic 3',5'-GMP was able to bind to the binding protein, but the affinity of this nucleotide for the binding protein was only about one hundredth of that shown by cyclic AMP.

The cyclic AMP-binding protein has been purified 100-fold by the use of precipitation at pH 4.8, calcium phosphate gel adsorption, DEAE-cellulose chromatography followed by the application of a sucrose density gradient. It was possible to show that the cyclic AMP substrate for the binding protein was not modified as a result of adsorption to this protein [40]. This demonstrated that the nucleotide was not a substrate for a protein-enzyme. Hence it was possible that the role of the nucleotide was to modify the macromolecular structure of this cytosolic binding protein.

The presence of such a binding protein in the adrenal cortex with a high affinity and a high degree of specificity for cyclic AMP resulted in the use of this protein for the assay of cyclic AMP in tissue extracts [43].

This method allowed the determination of as little as 20 pmol of cyclic AMP in biological material, and the assay was based upon the use of isotopic dilution of added tritium-labelled cyclic AMP. The advantage of the assay was that it utilised an easily isolated protein from bovine adrenal glands, the tissue was readily accessible and the protein was reasonably stable.

4. Protein kinase

In the cytosol of the bovine adrenal cortex a protein kinase was identified which could be activated by cyclic AMP [42]. Studies of this kinase were performed using protamine, histones and other exogenous substrates as acceptors of the phosphate group, because the endogenous substrate of this protein kinase was not established. The adrenal cortex-activatable protein kinase was shown to respond to cyclic AMP and this nucleotide showed a K_m for the activation process of 1.4 × 10^{-8} M while the maximal stimulation with this nucleotide was accomplished at 1×10^{-7} M. In this nucleotide-stimulated reaction, the activity of the protein kinase was increased about 4-fold by the cyclic nucleotide [42].

Cyclic AMP was a much more effective activator of this cytosolic protein kinase than any of the other nucleotides studied, including cyclic IMP, cyclic CMP and cyclic GMP. These other cyclic nucleotides could mimic the effect of cyclic AMP when the concentration of the nucleotides was raised to the supraphysiological levels of 2×10^{-4} M [42,44].

Attempts have been made to try to isolate this cyclic AMP-activatable protein kinase from the bovine adrenal cortex cytosol fraction, and to separate this enzyme from the previously identified cyclic AMP-binding protein. By the use of ammonium sulphate precipitation, calcium phosphate gel adsorption, DEAE-cellulose chromatography and sucrose density gradient methods, it was possible to show that the cyclic AMP-binding proteins could be partially separated from the protein kinse [42]. It was clear, nevertheless, that the cyclic AMP-dependent protein kinase isolated from the bovine adrenal cortex retained significant cyclic AMP-binding activity even after extensive purification. The protein kinase activities and the cyclic AMP binding properties appeared to be enriched in parallel, although there was evidence that the binding protein was more labile to heating than the protein kinase.

It became clear that the cyclic AMP-binding protein was intimately associated with the protein kinase and, as a result of the association of the binding protein with the protein kinase, the activity of the protein kinase was suppressed [41]. A theory was advanced that the addition of cyclic AMP to such a system resulted in the binding of the cyclic AMP to the binding protein subunit, so that the suppression of the protein kinase activity by the binding protein was released [41,44]. Thus activation of this protein kinase could result from the dissociation of the binding protein from the protein kinase.

The cyclic AMP-binding protein was subsequently referred to as the subunit 'receptor', and the protein kinase was referred to as a subunit containing the 'catalytic' component of the protein kinase. These two subunits appear to be present in a complex as shown by ultracentrifugal and electrophoretic studies. Addition of cyclic AMP to the complex results in the cyclic AMP binding to the 'receptor' which, in turn, causes the latter to dissociate from the protein kinase. Consequently the liberated kinase is catalytically more active when freed from the receptor molecule.

Such a theory is consistent with the evidence that the diverse manifestations of cyclic AMP in different tissues result from tissue-specific protein or proteins phosphorylated by ATP in the presence of the activated protein kinase of the tissue. There is, therefore, a need to identify which protein or proteins in adrenal cortical cells are phosphorylated in the chain of events linking the interaction of ACTH with the plasma membrane of the cells to the ultimate secretion of corticosterone or cortisol by the adrenal cells.

Garren and associates showed that in adrenal tissue there was a cyclic AMP-dependent protein kinase capable of phosphorylating ribosomes [44]. The concentration of cyclic AMP needed to activate this protein kinase was within the physiological range of $10^{-8}-10^{-6}$ mol/l. On the assumption that the cyclic AMP-dependent protein kinase was playing a key role in the response of the adrenal cortex to ACTH activation, and hence to elevated cyclic AMP concentrations, theories were advanced to link the observed requirement for protein synthesis associated with the mode of action of ACTH with this protein kinase reaction. It was suggested that the cyclic AMP-dependent protein kinase might be involved in the phosphorylation of ribosomes which, in turn, might affect the translation of messenger RNA [44].

It has been shown that in the bovine adrenal cortex there is also a cyclic AMP-independent protein kinase. This protein kinase is termed a type III cyclic AMP-independent protein kinase; casein is an effective phosphate acceptor for this type III enzyme and is better than the usual substrate, histone. The molecular weight of this cyclic AMP-independent protein kinase in the adrenal cortex is greater than 300 000 [45]. The adrenal cortex cytosol also contains a cyclic AMP independent protein kinase which can utilise either ATP or GTP as the phosphate donor [46].

5. Proteins phosphorylated in the adrenal cortex

The widespread distribution of cyclic AMP-dependent protein kinases has suggested that such protein kinases may mediate all the actions of cyclic nucleotides in living systems. Such a concept implies that important control events in living cells are exerted through phosphorylation reactions. A considerable number of enzymes and proteins are known to be modified by phosphorylation as clearly demonstrated by the subject matter of this book. See also Reviews [47,48].

It has been shown that there are heat-stable inhibitor proteins in the adrenal cortex, and one of these is contaminated with an endogenous phosphate acceptor protein. This latter protein can be phosphorylated by protein kinase II and protein kinase III [48]. The possible role of protein kinase inhibitors will be dealt with later.

In attempting to interpret how cyclic AMP may modify production of

102

steroid hormones via protein phosphorylation events, none of the enzymes already known to be modified by phosphorylation was a candidate as an important rate-limiting or key controlling enzyme in steroidogenesis. This prompted our studies on the following enzyme system [49].

6. Cholesterol ester hydrolase

It is known that when animals are subjected to stressful situations, or when ACTH is administered to such animals, that there is a reduction in the cholesterol content of the adrenal cortex [50–52]. This drop in cholesterol content is almost entirely due to a decrease in the esterified cholesterol of the adrenal cells [52]. The cytosol of adrenal cells contains large quantities of esterified cholesterol found in association with triacylglycerol and phospholipids as lipid droplets [53]. In these cells the cholesterol esters appear to be mobilised as a result of ACTH stimulation of the adrenal cells. The administration of ACTH to rats followed by removal of the adrenals 15 min later resulted in activation of cholesterol ester hydrolase as measured in a crude adrenal homogenate [54].

Cholesterol ester hydrolase activity has been examined in different cell fractions from the adrenal cortex. It has been shown that there is cholesterol ester hydrolase activity in the mitochondrial fractions, in lysosomal fractions, in the endoplasmic reticulum and in the cell cytosol [53,54]. Quantitation of the cholesterol ester hydrolase activity in these different cell fractions expressed as enzymic activity/mg protein is shown in Table 1. The cholesterol ester hydrolase activity in the cell cytosol was of interest, because this is the fluid bathing the cholesterol ester-laden lipid droplets which are characteristic of rat and human adrenal cells.

Ether anaesthesia stress, which is known to increase adrenocorticotrophic hormone concentration in blood [55,56], has been shown to produce choles-

TABLE 1

The activity of cholesterol ester hydrolase determined at pH 7.4 in different adrenal cell fractions

Fraction	Cholesterol ester hydrolase % of total activity
Nuclei	2
Mitochondria	3.7
Lysosomes	11
Microsomes	18
Cytosol	67

terol ester depletion in the adrenals and an increase in cholesterol esterase activity in rat adrenal glands [49,57]. Cycloheximide administration to rats did not alter significantly the concentration of cholesterol and cholesterol esters in the lipid droplets in the adrenals. Injection of cycloheximide to rats followed by ether anaesthesia resulted in an increase in free cholesterol with a concomitant decrease in cholesterol ester concentration within the adrenal lipid droplets [49,53]. Cholesterol esterase and protein kinase activities in the rat adrenal 100 000 × g supernatant fraction were significantly higher in animals subjected to stress by ether anaesthesia prior to killing, than in control non-stressed or quiescent animals. Injection of cycloheximide did not prevent the stress-induced enhancement of the activities of either enzyme [43,49]. It seemed possible therefore that the interaction of ACTH with rat adrenal cells activated the adrenal cortical protein kinase system and the cholesterol ester hydrolase in the adrenal cytosol [49].

In vitro studies showed that the crude rat adrenal cortical cytosolic cholesterol ester hydrolase activity was significantly enhanced by the addition of cyclic AMP + ATP in the presence of theophylline. It was shown that, for maximal stimulation of the cholesterol ester hydrolase, the rat adrenal cytosol required an active protein kinase as well as cyclic AMP and ATP.

7. Cholesterol ester hydrolase assay

In all studies on the measurement of cholesterol ester hydrolase activity (or triacylglycerol lipase activity), a major problem arises concerning the best method for the presentation of the substrates to the enzyme. In the early investigations [49,53,57], studies were made at relatively low substrate concentrations with a substrate presented to the enzyme from an acetone solution. The assay was based upon the method of Chen and Morin [58]. The reaction mixture contained 50 mM Tris/HCl buffer pH 7.4, 5 mM $MgCl_2$, 25 mM KCl, 30 μM [4-^{14}C]cholesterol oleate in 10 μl of acetone. In general, the assay mixture contained 1–2 mg protein/ml incubation mixture and the enzymic reaction was started by the addition of substrate. The incubation was conducted for 30 min at 37 °C and the reaction rate was linear from 0–30 min. The reaction was terminated by the addition of acetone–alcohol and the protein was precipitated, centrifuged and an aliquot of the supernatant removed for radioactive cholesterol assay. Carrier cholesterol was added to the supernatant to increase the concentration of free cholesterol. The free cholesterol was precipitated with digitonin and the cholesterol digitonide was 'washed in the usual way with acetone–ether and then with ether. The precipitate was dissolved in methanol and an aliquot taken for counting in a liquid scintillation counter.

There are alternative methods for the measurement of cholesterol ester hydrolase activities in tissues [59]. Instead of measuring the cholesterol released by cholesterol ester hydrolase from the substrate cholesterol oleate, it is possible and often more convenient to measure the free fatty acid released from the ester. In one method described by Steinberg and associates [60], the cholesterol ester hydrolase is measured in terms of the [1-^{14}C]oleic acid liberated from cholesterol [1-^{14}C]oleate. The substrate is prepared by the injection of the labelled cholesterol ester in an ethanolic solution into a solution containing 50 mM phosphate buffer and 0.5% bovine serum albumin. The final substrate concentration is 0.5 mM and the solution contains 3.2% ethanol. It has been shown that this concentration of alcohol does not affect the activatable cholesterol ester hydrolase. The polar nature of the liberated fatty acid allows the labelled free fatty acid released as a result of the enzymic reaction to be separated very easily from the labelled unreacted ester substrate [59].

This method has the additional advantage that it is equally applicable to esters of cholesterol and esters of glycerol. Thus, by this method, it is possible to contrast the cholesterol ester hydrolase activity and the triglyceride lipase activity using comparable methods and comparable substrate concentrations. This is important, as will be discussed later in considering the substrate specificity of the cholesterol ester hydrolase, because this enzyme is found in the adrenal cortex and in adipose tissue [53,54,57,59–67]. The adrenal cortex and adipose tissue also contain triacylglycerol lipase enzymes in the cytosol that are activatable by protein kinases [60,68–77]. The degree of activation of both cholesterol and glycerol ester hydrolases is similar; hence there is a need for careful comparison of the two activities in order to decide whether these two enzymes are a single entity or separate enzymes. The use of the labelled oleate release technique simplified the methodological aspects of the comparison of the two reactions.

8. Protein kinase assay

Protein kinase activity was determined by the method of DeLange and co-workers [78]. The assay system consisted of 50 mM sodium 3-glycerophosphate pH 6.5, 0.6 mM [γ-^{32}P]ATP sodium salt, 0.5 mM cyclic AMP, 5 mM MgCl$_2$, 4 mM theophylline, 20 mM NaF and histone, 2.4 mg/ml. The reaction mixture had a final volume of 0.125 ml. The incubation was conducted for 20 min at 30 °C along with the appropriate control. The enzyme unit was defined as micromoles of phosphate transferred from the gamma position of ATP onto the histone under the conditions employed. The phosphorylated protein was isolated as described by Corbin [79] dissolved in 23 M formic acid and an aliquot taken for counting by liquid scintillation spectrometry. The reaction rate was linear within the period of incubation and protein concentrations ·used.

9. Concentration of cyclic AMP required to activate the adrenal cortical cytosolic cholesterol ester hydrolase in the presence of protein kinase

It was necessary to determine the concentration of cyclic AMP required for the activation of the cholesterol ester hydrolase. In such experiments the rat adrenal cytosol, containing cholesterol ester hydrolase and protein kinase, was desalted on a Sephadex G-25 column and then incubated with varying concentrations of cyclic AMP in the presence of 1 mM ATP and theophylline for 30 min. The substrate [4-^{14}C]cholesterol oleate was added to the reaction mixture and the incubation continued for a further 30 min. The cholesterol ester hydrolase activity was then assayed as described previously. As shown in Figure 2, in the presence of 1 mM ATP, cholesterol ester hydrolase was activated by cyclic AMP in the presence of protein kinase. The apparent K_m for the cyclic AMP in this activation reaction was of the order of 10^{-7}M.

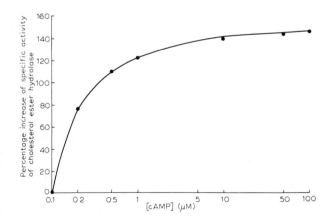

Figure 2. The activation of cholesterol ester hydroxylase obtained from the desalted adrenal cortical cytosol preparation. Each incubation contained enzyme plus 1 mM ATP and 1 mM Mg^{2+} in addition to the indicated amount of cAMP.

10. Concentration of ATP required to activate the cytosolic adrenal cortical cholesterol ester hydrolase

In a similar fashion, the activation of desalted rat adrenal cytosolic cholesterol ester hydrolase was investigated at a fixed concentration of cyclic AMP 100 times the K_m value, namely 10 μM, in the presence of theophylline, and the concentration of ATP in the reaction mixture was varied. The incubation period, as before, was 30 min at 37 °C. After this preincubation period the substrate [4-^{14}C]cholesterol oleate was added and the incubation continued for a further 30 min. The reaction was then stopped and the activity of cholesterol ester hydrolase determined, as described previously. As shown in Figure 3 the activation of cholesterol ester hydrolase in the presence of 10 μM cyclic AMP showed a K_m for ATP of the order of 10^{-5} M; thus, in subsequent studies to activate cholesterol ester hydrolase in the presence of the cyclic AMP-dependent protein kinase, we added 10 μM cyclic AMP, (100 times the K_m for this nucleotide) and 1–5 mM ATP, (100 times the ATP K_m value) were used.

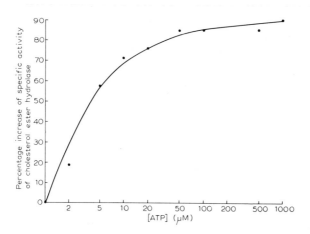

Figure 3. The activation of cholesterol ester hydrolase obtained from the desalted adrenal cortical cytosol preparation. Each incubation contained 10 μM cAMP and 1 mM Mg^{2+} in addition to the indicated amount of ATP.

11. Rate of activation of adrenal cortical cholesterol ester hydrolase

The time course of activation of cholesterol ester hydrolase was determined in a similar manner. The cytosol preparation of rat adrenal tissue was incubated with 10 μM cyclic AMP and 1.0 mM ATP in the presence of theophylline, and after varying periods of incubation at 37 °C the reaction mixture was diluted, desalted through a small Sephadex G-25 column and the activity of the cholesterol ester hydrolase determined in the usual way. The rate of activation of cholesterol ester hydrolase was extremely rapid, showing that a preincubation period of 20 min at 37°C achieved about 90% of the maximal activation of the enzyme [65,66]. The characteristics of this activation reaction are shown later in Figure 9.

12. The pH optimum of the adrenal cortical cholesterol ester hydrolase

In a similar fashion the optimum pH of the rat adrenal cytosolic cholesterol ester hydrolase was investigated, and it was shown that the enzyme had a fairly broad pH optimum between pH 7.3 and 7.9 as shown in Figure 4. In all subsequent studies, observations on cholesterol ester hydrolase activity in the rat adrenal cytosol preparation were made at pH 7.4

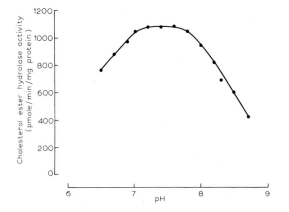

Figure 4. The effect of different hydrogen ion concentrations on the activity of the adrenal cortical cytosol cholesterol ester hydrolase.

13. Cholesterol ester hydrolase in the bovine adrenal cortex

These studies showed that rat adrenal cytosolic cholesterol ester hydrolase activity could be significantly enhanced in vitro by preincubation with cyclic

AMP and ATP in the presence of endogenous protein kinase [53,57]. The limitation imposed by the size of the rat adrenal glands required adrenals from larger animals to be used to permit a study of the mechanism of activation of adrenal cortical cholesterol ester hydrolase. Bovine adrenal glands were shown to contain a cytosolic cholesterol ester hydrolase which, like that of the rat, is activated by cyclic AMP + ATP [61,66].

Further studies on the bovine adrenal cortex cholesterol ester hydrolase showed that the preincubation of the cytosolic cholesterol ester hydrolase preparation in the presence of 5 mM $MgCl_2$ caused a marked decrease in cholesterol ester hydrolase activity. Under these conditions cyclic AMP and ATP protected the enzyme against such magnesium ion-dependent deactivation.

14. Methods for the purification of adrenal cortical cytosolic cholesterol ester hydrolase

Method 1 — The availability of bovine adrenal glands, which had been shown to contain an activatable cholesterol ester hydrolase in the cytosol of their cells, allowed attempts to be made to purify the hydrolase. In the first method [65] the adrenal cortical tissue is scraped off the capsule and a 20% homogenate of the tissue made in 0.25 M sucrose. The homogenate is centrifuged at 600 × g for 10 min to remove the cell debris, and the supernatant is centrifuged at 10 000 × g for 15 min to remove mitochondria. The supernatant from this procedure is centrifuged at 100 000 × g for 60 min to sediment the endoplasmic reticulum, giving an infranatant and a floating lipid layer. The lipid layer consisting of lipid droplets is carefully removed from the top of the tube, and the infranatant retained as the source of the enzyme. To this enzyme solution 1.0 M phosphate buffer pH 7.4 is added to make the final concentration of phosphate 0.10 M.

The hydrolase enzyme in sucrose and phosphate buffer is treated with solid ammonium sulphate slowly and with stirring at 0 °C to give a final concentration of 40% ammonium sulphate. The solution is centrifuged at 10 000 × g for 10 min. The protein pellet is carefully collected and dissolved in 20 mM phosphate buffer pH 7.4 and desalted by passage through a Sephadex G-25 column to remove ammonium sulphate. It has been shown that almost all of the cytosolic cholesterol ester hydrolase is precipitated in the ammonium sulphate procedure between the concentration 0–40% ammonium sulphate.

The desalted protein fraction is applied to a Sephadex G-200 column previously equilibrated with 20 mM phosphate pH 7.4, and the protein eluted with the same buffer. Fractions from the column are assayed for cholesterol ester hydrolase and for protein. As shown in Figure 5, most of the cholesterol ester hydrolase is eluted in the void volume of the Sephadex G-200 column and a smaller peak is eluted with retention characteristics about twice the void volume. The ratio of the amounts of the two cholesterol ester hydrolases varies from preparation to preparation. The void volume fraction containing most of the cholesterol ester hydrolase is concentrated and used in the next stage of the purification.

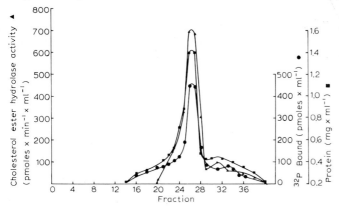

Figure 5. The elution profile from Sephadex G-200 of an adrenal cortical cytosol cholesterol ester hydrolase. For details see text.

The protein from the void volume fraction from the Sephadex G-200 column is applied to a Sepharose 4B column which has been equilibrated with 20 mM phosphate buffer pH 7.4, and the protein is eluted with the same buffer. Fractions are collected and assayed for cholesterol ester hydrolase and for protein. The results of this procedure are shown in Figure 6. Again the major peak is in the void volume. This fraction is usually turbid, due to the presence of lipid which could be partly removed from the preparations by high speed centrifugation or by hydroxyl apatite chromatography.

The protein fraction from the void volume of the Sepharose 4B column is concentrated and applied to a hydroxyl apatite column. The protein is eluted from the column with stepwise elutions of two and a half column volumes of potassium phosphate buffer pH 7.4. The protein eluted at 400 to 500 mM phosphate buffer is pooled and desalted on a Sephadex G-25 column equilibrated with 50 mM potassium phosphate buffer and eluted with the same buffer. Although the hydroxyl apatite step does not increase the specific

110

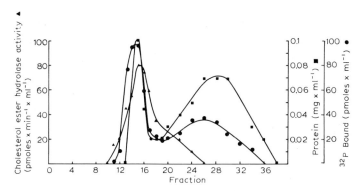

Figure 6. The elution profile from Sepharose 4β of an adrenal cortical cytosol cholesterol ester hydrolase. For details see text.

activity of the cholesterol ester hydrolase it removes almost all of the protein kinase in the preparation.

The steps used in this Method 1 for the purification of the cholesterol ester hydrolase achieve a 57-fold purification of the enzyme.

Method 2 — This purification of cholesterol ester hydrolase is adapted from published methods [59,70]. In this technique the adrenal cortical tissue is homogenised in 0.25 M sucrose containing 10 mM phosphate buffer pH 7.4 containing 1 mM EDTA. The homogenate is centrifuged at $10\,000 \times g$ for 20 min, the floating lipid layer is carefully removed and discarded, and the infranatant is decanted from the pellet, the latter being discarded. The infranatant is centrifuged at $100\,000 \times g$ for 60 min; again the lipid layer is carefully removed and discarded, while the infranatant is decanted from the pellet and the latter is discarded.

The infranatant is carefully adjusted to pH 5.0 using acetic acid, care being taken to keep the temperature at 4°C. The mixture is centrifuged at $10\,000 \times g$ for 20 min and the pellet retained. The protein pellet is dissolved in 20 mM phosphate buffer pH 7.4 and centrifuged at $100\,000 \times g$ for one hour. The lipid layer on the top of the tube is carefully removed and discarded, and the infranatant is decanted from the pellet which is discarded. The infranatant is concentrated by ultrafiltration and the solution is placed on a Biogel A 150 column which is eluted with 20 mM phosphate buffer pH 7.4.

The enzyme is eluted from this column in the void volume; consequently the void volume fraction is pooled and concentrated by ultrafiltration. The protein solution is layered onto a preformed sucrose gradient and centrifuged using a swing-out rotor at $100\,000 \times g$ for 46 h. The tube is pierced and the different fractions carefully removed from the tube. Figure 7 shows the results

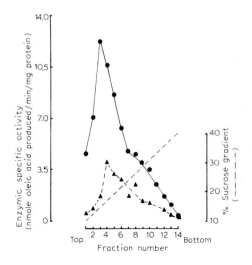

Figure 7. The distribution of cholesterol ester hydrolase (● —— ●) and triacyl-glycerol lipase (▲---▲), in fractions obtained after centrifugation in a 10-40% linear sucrose density gradient.

of the distribution of the cholesterol ester hydrolase activity on a typical sucrose density gradient separation. Table 2 shows the specific activity of the cholesterol ester hydrolase achieved at different stages of the purification procedure.

Partially purified cholesterol ester hydrolase, free from protein kinase activity could not be activated by cyclic AMP in the presence of ATP. This partially purified cholesterol ester hydrolase could, however, be activated by 10 μM cyclic AMP and 5 mM ATP in the presence of protein kinase. The

TABLE 2

Specific activity and enzymic recovery of cholesterol ester hydrolase during various steps in enzyme purification

Fraction	Cholesterol ester hydrolase		
	Enzymic specific activity (pmole oleic acid/ min/mg protein)	Total enzymic recovery (%)	Apparent purification
Adrenal cytosol	142	100	1
pH 5.0 precipitate	368	50	2.6
Bio-gel A fraction	705	10	5.0
Sucrose gradient fraction	3060	7.5	21.5

partially purified cholesterol ester hydrolase could be activated by cyclic AMP independent protein kinase + ATP in the absence of cyclic AMP. These results are shown in Table 3.

A procedure for the purification of a radioactively labelled cholesterol ester hydrolase from bovine adrenal cortical 100 000 × g supernatant was established [65]. This procedure involved preincubation of a crude enzyme extract with [γ-^{32}P]ATP followed by purification steps as described previously, which resulted in the isolation of a phosphorylated preparation of cholesterol ester hydrolase.

TABLE 3

The effects of nucleotides, protein kinase and protein kinase inhibitor on crude and partially purified bovine adrenal cholesterol ester hydrolase

Addition	Crude bovine adrenal cholesterol ester hydrolase (%)	Purified bovine adrenal cholesterol ester hydrolase (%)
None	100	100
ATP 5 mM	172	–
cAMP 10 μM	111	94
ATP + cAMP	250	109
Protein kinase inhibitor (PKI)	106	
ATP + cAMP + PKI	117	
ATP + cAMP + protein kinase		182

Bovine adrenal cortical 100 000 × g supernatant was prepared and potassium phosphate buffer added to give a final concentration of 0.1 M. Solid ammonium sulphate was added to this solution to produce 40% saturation. The solution was centrifuged at 10 000 × g for 10 min. The supernatant was decanted and the pellet redissolved in 20 mM potassium phosphate buffer pH 7.4. The passage of the protein solution through a Sephadex G-25 column removed the ammonium sulphate and endogenous nucleotides. It was possible to label the enzyme with [γ-^{32}P]ATP using the pooled desalted protein fractions from the Sephadex G-25 column. This protein fraction, free from nucleotides, was incubated at 30°C for 20 min with 10 μM cyclic AMP, 5 mM ATP containing 250 μCi of [γ-^{32}P]ATP, 1 mM MgCl$_2$, 4 mM NaF and 2 mM theophylline. After the preincubation, the protein solution was again desalted by passage through a Sephadex G-25 column equilibrated with 20 mM potassium phosphate buffer pH 7.4 [65]. The pooled protein fractions from the Sephadex G-25 column were applied to a Sephadex G-200 column and equilibrated with 20 mM potassium phosphate buffer pH 7.4. Most of the cholesterol ester hydrolase was eluted in the void volume fraction from the Sephadex G-200 column. The partial purification of the enzyme was conducted as previously described.

Fractions from the Sephadex G-200 and Sepharose 4B columns were assayed for cholesterol ester hydrolase activity and protein-bound ^{32}P radio-activity. The elution profiles obtained from the Sepharose 4B columns are shown in Figure 8. After protein precipitation with trichloroacetic acid, ^{32}P radioactivity remained bound to the protein fractions that contained cholesterol ester hydrolase activity. However, other proteins eluted after the void volume fractions were also found to be phosphorylated, but did not exhibit cholesterol ester hydrolase activity.

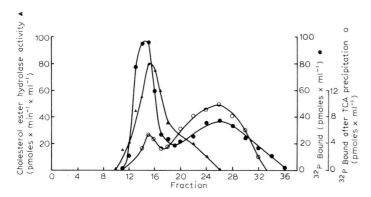

Figure 8. The elution profile from Sepharose 4β of an activated and phosphorylated cholesterol ester hydrolase preparation. For details see text.

Protein in the void volume from the Sepharose 4B column was concentrated by freeze-drying and subjected to gel electrophoresis on 5% polyacrylamide gel. On staining for protein and esterase activity, two protein bands were observed and the protein with the higher electrophoretic mobility exhibited esterase activity. When the gel was sliced and counted for ^{32}P radioactivity the isotope was mainly in the protein which exhibited esterase activity.

In other experiments, the phosphorylated protein fraction from the Sepharose 4B columns was subjected to electrophoresis on a dodecyl-sulphate gel and, on staining the gel, two proteins were found in close proximity to one another. When the gel was sliced and counted for radioactivity the protein band of greater electrophoretic mobility contained the radioactive isotope.

When the Sepharose 4B void volume fraction was examined using

114

electrophoresis on dodecyl sulphate gels with bovine serum albumin, immuno-globulin, ovalbumin and myoglobin as markers, the subunit molecular weight of cholesterol ester hydrolase was found to be about 41 000 ± 280.

The time course of activation and phosphorylation of cholesterol ester hydrolase by cyclic AMP-dependent protein kinase was investigated [65]. The estimation of activation and phosphorylation of the enzyme over a 20 min incubation period was carried out on the purified preparation of the enzyme obtained from the void volume fraction of the Sepharose 4B chromatography. In this instance, protein obtained from the ammonium sulphate precipitate had been incubated with 5 mM $MgCl_2$ for 10 min at 30 °C. This method was found to give a preparation of cholesterol ester hydrolase which had, as far as possible, been deactivated and dephosphorylated. Each incubation contained 10 ml of purified deactivated cholesterol ester hydrolase, beef heart cyclic AMP-dependent protein kinase, 10 μM cyclic AMP, 1 mM $MgCl_2$ and 30 mM KCl. One incubation was started by the addition of 20 μl of 5 mM ATP, and the other incubation by the addition of 20 μl of 5 mM ATP containing 0.1 mCi [γ-^{32}P]ATP. 1 ml samples were taken from the incubation containing [^{32}P]ATP and assayed for protein-bound ^{32}P radioactivity. At the same time 1 ml samples were taken from the duplicate incubation into EDTA at a final concentration of 4 mM to stop phosphorylation. Cholesterol ester hydrolase activity was then assayed as before. As shown in Figure 9 after 15 min, maximal activation of cholesterol ester hydrolase was achieved. For the first 4 min, phosphorylation closely paralleled cholesterol ester hydrolase activity. When cholesterol ester hydrolase activity reached a plateau at 15 min, a slow

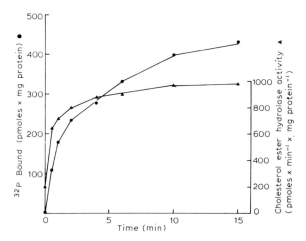

Figure 9. The time course of activation and phosphorylation of a cholesterol ester hydrolase preparation from the bovine adrenal cortex. For details see text.

incorporation of ^{32}P radioactivity into protein continued. The characteristics
of this rate of activation and phosphorylation were similar to those of the
rates of activation observed previously.

A study was made of the effect of cyclic AMP-dependent protein kinase
inhibitor on cholesterol ester hydrolase activity. The addition of 5 mM ATP
and 5 mM MgCl$_2$ to a desalted bovine adrenal cortical 100 000 × g
supernatant preparation gave a marked stimulation of cholesterol ester
hydrolase activity. The addition of 10 μM cyclic AMP with 5 mM ATP and 5
mM MgCl$_2$ produced an even greater stimulation. When cyclic AMP-
dependent protein kinase inhibitor was added to the 105 000 × g desalted
supernatant at a final concentration of 100 μg inhibitor/ml it had no effect on
cholesterol ester hydrolase activity, but when it was added prior to the
addition of cofactors ATP and cyclic AMP, stimulation was completely
abolished.

15. Deactivation of cholesterol ester hydrolase by magnesium ions

The effect of magnesium ions on cholesterol ester hydrolase activity was
examined. It was known that magnesium ions promoted the deactivation of
cholesterol ester hydrolase. Consequently, the effect of magnesium ions on
the amount of protein bound ^{32}P radioactivity was investigated [65]. A
partially purified preparation of cholesterol ester hydrolase obtained from
Sephadex G-200 chromatography was phosphorylated by preincubation of the
void volume fraction at 30°C for 20 min with 10 μM cyclic AMP, 1 mM
MgCl$_2$, 30 mM potassium chloride, 20 mM potassium phosphate buffer, pH
7.4 and 1.0 mM ATP containing 125 μCi of [γ-^{32}P]ATP. The mixture, after
incubation, was desalted by passage through a Sephadex G-25 column equilibrated with 10 mM potassium phosphate buffer pH 7.4. Desalted ^{32}P
phosphorylated preparation was then incubated with 5 mM magnesium
chloride. As shown in Figure 10 deactivation of cholesterol ester hydrolase
closely paralleled the dephosphorylation as judged by the loss of ^{32}P from the
protein fraction.

16. Deactivation of cholesterol ester hydrolase by a phosphoprotein phosphatase

Evidence has been presented to support the concept that in the rat and bovine

116

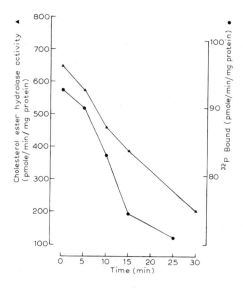

Figure 10. The effect of Mg^{2+} ions on the dephosphorylation and deactivation of cholesterol ester hydrolase from the bovine adrenal cortex.

adrenal cortex the cytosolic cholesterol ester hydrolase enzyme is activated by a covalent phosphorylation of the enzyme. The phosphorylation of this enzyme occurs by the transference of the γ-phosphate of ATP to the inactive cholesterol ester hydrolase under the influence of a cyclic AMP-dependent protein kinase. It is known that phosphorylase kinase is activated by a similar mechanism [47] and this enzyme in its active form is deactivated by a phosphoprotein phosphatase [47,80]. The crude, active, cytosolic cholesterol ester hydrolase enzyme from the bovine adrenal cortex is deactivated by millimolar concentrations of magnesium or calcium ions [65,81]. By contrast magnesium and calcium ions cannot deactivate or inhibit the partially purified enzyme suggesting that in the purification of the hydrolase the enzyme preparation lost some other factor which was instrumental in deactivating the hydrolase in the presence of calcium or magnesium ions. It is possible that the magnesium or calcium ion-dependent factor present in the bovine adrenal cortical cytosol was a phosphoprotein phosphatase [64,65].

The addition of 5 mM $MgCl_2$ to a desalted bovine adrenal cortical cytosolic preparation followed by incubation at 37 °C resulted in a rapid deactivation of the hydrolase to a value about 5% of the control preincubation value. The addition of the same concentration of $MgCl_2$ to a similar desalted cytosolic preparation in the presence of 5 mM ATP and 10 μM cyclic AMP protected the hydrolase against this deactivation reaction.

To explore further the role of the phosphoprotein phosphatase in the modulation of the bovine adrenal cortical cytosolic cholesterol ester hydrolase the enzyme was partially purified as detailed previously. The protein component in the void volume fraction from the Sepharose 4B column was centrifuged at 100 000 × g for 30 min to remove the turbidity. The floating lipid layer was carefully removed and the infranatant was used as the source of the partially purified cholesterol ester hydrolase.

The hydrolase was activated by preincubation for 10 min with 50 mM sodium glycerophosphate, 1 mM MgCl$_2$, 4 mM NaF, 2 mM theophylline, 10 μM cyclic AMP and 1 mM ATP. Upon completion of the preincubation period the mixture was desalted on a Sephadex G25 column and the eluate adjusted to contain 1 mM EDTA. This phosphorylated and active enzyme was then incubated at 37 °C with protein phosphorylase-1 from skeletal muscle (see Chapter 1) in the presence or in the absence of manganous ions. The result is shown in Figure 11. Thus this phosphoprotein phosphatase is capable of deactivating and dephosphorylating the enzyme and the dephosphorylation is influenced by manganous ions. This result was subsequently confirmed employing an activated cholesterol hydrolase labelled with ^{32}P from [γ-^{32}P]ATP. The loss of ^{32}P from the active enzyme correlated with the loss of enzymic activity. However, the number of phosphorylation sites of this cholesterol ester hydrolase have yet to be discovered.

In summary, the active phosphorylated form of cholesterol esterase can be dephosphorylated and deactivated by a divalent cation dependent

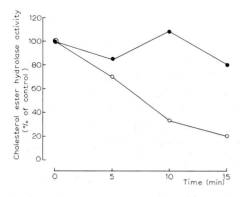

Figure 11. The effect of incubation with a phosphoprotein phosphatase on the deactivation of cholesterol ester hydrolase. (●——●) Phosphorylase kinase phosphatase; (○——○) Phosphorylase kinase phosphatase + Mn^{2+}. For details see text.

phosphoprotein phosphatase. The activity of this phosphoprotein phosphatase can be moderated by ATP and is possibly the same enzyme isolated some time ago from the adrenal cortex [82].

17. Possible control of a phosphoprotein phosphatase in the adrenal cortex

Considerable evidence has now accrued suggesting that in the adrenal cortex and in adipose tissue the activity of cholesterol ester hydrolase and triacylglycerol lipase is dependent upon a balance between the phosphorylated or active form of the lipolytic enzymes and a dephosphorylated or inactive form of the enzymes [65,75]. This balance being achieved in part through the action of cyclic AMP dependent protein kinases and ATP. The possibility also exists that the balance between the active and inactive form of the enzymes could be modulated through changes in the activity of phosphoprotein phosphatases. Unfortunately it is not yet known whether the phosphoprotein phosphatase present in the adrenal cortex or in adipose tissue is specific for cholesterol ester hydrolase under consideration in this report or alternatively whether the phosphoprotein phosphatase is an enzyme with a broad specificity.

An additional complexity in attempts to isolate, characterise and study phosphoprotein phosphatases arises due to the fact that other proteins or peptides present in mammalian tissues are capable of modifying the activity of phosphoprotein phosphatases. A heat stable inhibitor of phosphorylase phosphatase has been isolated from rat skeletal muscle, rat liver, beef heart and beef adrenal cortex [83]. There seems to be at least two types of phosphoprotein phosphatase inhibitor which have been separated using DEAE-cellulose chromatography [83]. The inhibitors have been classified as Type I and Type II. The activity of the Type I phosphoprotein phosphatase inhibitor is regulated by phosphorylation. This phosphorylatable inhibitor is present in the adrenal cortex. Under physiological circumstances there is therefore in the adrenal cortex a peptide capable of inhibiting phosphoprotein phosphatase and the efficacy of this inhibitor is increased by phosphorylation. The phosphorylation of this inhibitor is catalysed by cyclic AMP-dependent protein kinase. The deactivation of cholesterol ester hydrolase is deduced to be under the influence of a phosphoprotein phosphatase. If this latter phosphoprotein phosphatase is controlled by a phosphorylatable inhibitor then the conversion of cholesterol ester hydrolase to its active form and the retention of the enzyme in its active form will be facilitated by the activation of the inhibitor of the phosphoprotein phosphatase. It is possible that the phosphorylation of an inhibitor of phosphoprotein phosphatase by cyclic AMP-dependent protein kinase is an important factor in the control of the activity of cholesterol ester hydrolase in the cytosol of the adrenal cortex under physiological conditions. In this respect the situation may be similar to glycogen metabolism (see Chapter 1).

Other studies on the adrenal cortex have emphasised that one phosphoprotein phosphatase, the phosphorylase phosphatase may exist in two different forms. One type of phosphorylase phosphatase is spontaneously active, it has a molecular weight of 95 000 and is inhibited by ATP-Mg^{2+}. This ATP-Mg^{2+} inhibition is freely reversible and there is apparently no covalent modification to the enzyme. In the other form phosphorylase phosphatase is found to be ATP-Mg^{2+}-dependent and has a molecular weight of about 88 000, again the activation of this enzyme is also reversible. The ATP-Mg^{2+}-dependent phosphorylase phosphatase is activated by Mg^{2+} irreversibly, the optimal Mg^{2+} concentrations being about 4 mM [84]. The stimulation of the ATP-Mg^{2+}-dependent phosphoprotein phosphatase is specific for ATP and Mg^{2+}, and this reaction does not appear to be influenced by a kinase. This latter activation of phosphoprotein phosphatase does not involve a covalent modification of the enzyme but the ATP-Mg^{2+} activation involves a relatively tight binding of the nucleotide and Mg^{2+} to the enzyme. It is not known whether either of these enzymes are capable of dephosphorylating cholesterol ester hydrolase and triacyl glycerol lipase under consideration in this paper [84].

18. Role of calcium ions in steroidogenesis

There has been considerable interest in the possible role of calcium ions as secondary messengers in certain tissues [6,85,86]. It has been established that the response of rat adrenal cortical cells to ACTH in vitro is dependent on the calcium ion concentration [87]. Furthermore the rat adrenal mitochondrial cholesterol side chain cleavage reaction is activated by calcium ions, in the presence of the substrate cholesterol [88,89]. However the concentration of calcium ions required to produce this effect is sufficient to completely uncouple the mitochondria [89]. Despite this objection there is considerable interest in the possible role of calcium ions as intracellular modulators of various metabolic reactions including steroidogenesis.

It is established that certain divalent ions influence the activity of cholesterol ester hydrolase, and it is possible that in adrenal cortical cells there may be a calcium-dependent modulator protein akin to that described in muscle and other tissues [80,90,91]. This modulator protein is considered to influence adenyl cyclase, the calcium-dependent ATPase, the red blood cell calcium ion transport process and so on.

Calcium ions do influence the adrenal cortical cytosolic cholesterol ester hydrolase but the exact mode of action of calcium ions in this complex process has not been established [65].

While the effect of calcium ions on cholesterol ester hydrolase in the adrenal cortex is still unclear it has been shown that calcium ions increase prostaglandin E and $F_{2\alpha}$ production in cat adrenals in vivo. In the view of certain workers [28] the interaction of ACTH with its receptor in the plasma membrane may result in activation of a phospholipase A_2. The predominant free fatty acid released is arachidonate from the C_2 position of the phospholipid and this fatty acid by the prostaglandin synthetase system yields prostaglandin E_2 and $F_{2\alpha}$. These prostaglandins in turn could act as sophisticated calcium ion ionophores and alter steroidogenesis and steroid secretion by perturbing the free calcium ion concentration in the cell [28]. There are many other aspects of steroid hormone synthesis and secretion which could be influenced by calcium ions.

19. Characteristics and substrate specificity of the cholesterol ester hydrolase in the adrenal cortex

In the lipid droplets of the adrenal cortex there is a variable quantity of esterified cholesterol as well as triglyceride and a substantial amount of phospholipid. Studies have been made on the adrenal cortical cytosol to try to establish whether the cholesterol ester hydrolase present in that cell fraction has a selective action on specific esters.

The natural substrate for the adrenal cortical cytosolic cholesterol ester hydrolase is the esterified cholesterol contained in the lipid droplets in the cytoplasm. As discussed previously these lipid droplets contain triglycerides and phospholipids as well as cholesterol esters. Cholesterol is esterified to a range of long chain fatty acids; the analysis of rat adrenal cortical cholesterol esters is shown in Figure 12. The principal ester is cholesterol oleate and for this reason many investigators tend to use radioactively-labelled cholesterol oleate as the substrate in assays of cholesterol ester hydrolase. Nevertheless it is important to investigate whether the activatable adrenal cortical cholesterol ester hydrolase can discriminate between the various esters contained within the lipid droplets. The activity of rat adrenal cytosolic cholesterol ester

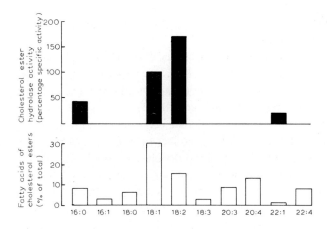

Figure 12. Fatty acids composition of rat adrenal cholesterol esters (unshaded). The activity of rat adrenocortical cytosolic cholesterol ester hydrolase (shaded) against different cholesterol esters of rat adrenal lipid droplets was measured as the percentage release of free fatty acids. See text for details.

hydrolase was measured against a range of different cholesterol esters and the results are shown in Table 4. These results demonstrate that this enzyme has an apparently higher activity towards the unsaturated fatty acid esters the activity against cholesterol erucate being much less. One of the problems associated with a comparison of the activity of this enzyme towards these different esters is that the esters have quite different physical characteristics, such as the melting point. Thus the particle size, and hence the surface area, of the substrates will differ even when apparently identical methods are used for the preparation of the substrate dispersions. Furthermore in the naturally occurring substrate particles there are phospholipids and triglycerides present

TABLE 4

The effect on cholesterol ester hydrolase activity of altering the characteristics of the sterol ester substrate. The change in fatty acid degree of desaturation and the change in the sterol side chain on the hydrolase activity

Substrate	Cholesterol ester hydrolase (%) specific activity
Cholesteryl oleate	100
Norcholesteryl oleate	154
Pregnenol oleate	123
Cholesteryl linoleate	150

and it is known that the latter is also a possible substrate for this enzyme as will be discussed in another section.

It is possible to approach the substrate specificity of the activatable cholesterol ester hydrolase in a different way. The lipid droplets from the adrenal cortex can be resuspended in an adrenal cortical cytoplasmic preparation containing the cholesterol ester hydrolase and the latter enzyme may be activated with cyclic AMP and ATP in the usual way. Incubation of such a mixture followed by the isolation of the released free fatty acids confirms that the principal fatty acids released from the lipid droplets are linoleate, arachidonate and oleate with much smaller amounts of palmitate and stearate (Figure 12). It has been shown that in the adrenal cytosol, ATP, cyclic AMP and a cyclic AMP-dependent protein kinase activate cholesterol ester hydrolase by almost 40% while the triacylglycerol lipase is only activated by about 12%. In this study [60] it was concluded that the two enzymes were different on the basis of a differential inhibition using an organophosphate, chloropyrifos oxone. Cholesterol ester hydrolase activity was more sensitive to inhibition by chloropyrifos oxone than was triacylglycerol lipase. In five experiments, 50% inhibition of triacylglycerol lipase activity was obtained at concentrations of the inhibitor 50 to 90 times the concentration required to effect 50% inhibition of cholesterol ester hydrolase. These workers obtained further evidence supporting the concept that the two enzymic activities may reside in different enzymes from ammonium sulphate fractionation of the adrenal cytosol fraction. The peak cholesterol ester hydrolase activity occurred in the precipitate at 20 to 30% ammonium sulphate saturation while the peak triacylglycerol lipase activity occurred at the 30 to 40% ammonium sulphate saturation. In the laboratory of the authors of this review using the bovine adrenal cortex it has not been possible to obtain such clear cut separations of cholesterol ester hydrolase activity from triacylglycerol lipase activity using the ammonium sulphate fractionation scheme. Similarly the inhibitor chloropyrifos oxone does not in our hands prove to be as selective on cholesterol ester hydrolase compared to triacylglycerol lipase. This could be due to different methods used in the preparation of the cholesterol ester hydrolase.

The adrenal cytosol contains triacylglycerol as well as esterified cholesterol and the cytosol contains activatable triacylglycerol lipase and activatable cholesterol ester hydrolase [60,69]. Attempts have been made to establish whether triacylglycerol lipase and cholesterol ester hydrolase behave kinetically as a single enzyme or as distinct activities. Both enzyme activities were assayed against a common substrate made of equimolar portions of cholesterol oleate and trioleylglycerol sonicated together. Mixed emulsions of cholesterol oleate and trioleylglycerol were prepared by sonicating together either labelled cholesterol oleate and unlabelled trioleylglycerol or unlabelled cholesterol oleate and labelled trioleylglycerol. In each case the concentration of cholesterol oleate and trioleylglycerol was 3 mM. It was shown that unlabelled cholesterol oleate inhibited the hydrolysis of labelled cholesterol oleate and inhibited the hydrolysis of trioleylglycerol. Conversely the unlabelled trioleylglycerol inhibited the hydrolysis of cholesterol oleate. In each case the percentage inhibition was similar. The authors interpreted these results as showing that these two activities bind to the substrate emulsions in quantitatively similar fashions supporting the concept that the two activities bind in concert. From these results it was concluded that the enzymes are very similar in their ability to bind to emulsions or alternatively that they may bind as a unit.

It is known that the mode of presentation of hydrophobic substrates to enzymes that act at interfaces is of crucial importance. The nature and the size of the emulsion particles as well as the presence or absence of emulsifiers influences the kinetics of enzymes such as triacylglycerol lipase and cholesterol ester hydrolase [92].

20. Adipose tissue

Rat adipose tissue contains an activatable triacylglycerol lipase and an activatable cholesterol ester hydrolase [59]. In isolated rat adipocytes almost all of the triacylglycerol lipase and the cholesterol ester hydrolase activities were found in the 100 000 × g supernatant fraction showing that the distribution of this enzyme in the adipocytes is similar to that found in the adrenal cortex [59]. The activation of the cholesterol ester hydrolase in adipocytes is dependent upon protein kinase because the presence of the protein kinase inhibitor from muscle blocks the ATP inactivation of the ester hydrolase. In intact adipocytes epinephrine was also able to activate the esterase suggesting that the cholesterol ester hydrolase in adipose tissue was activated by similar physiological signals to those which activate the triacylglycerol lipase enzyme [59]. The exact significance of cholesterol ester hydrolase is adipose tissue is not clear. Attempts to isolate the triacylglycerol lipase enzyme from adipose tissue show that the cholesterol ester hydrolase appears to cofractionate with the triacylglycerol lipase [59].

Adipose tissue is important in lipid dynamics in mammals. This tissue receives free fatty acids and cholesterol through the action of lipoprotein lipase on chylomicrons and very low density lipoproteins. This tissue can produce triglycerides and cholesterol esters from these precursors and has an activatable triacylglycerol lipase and cholesterol ester hydrolase. The former activity is very important in effecting the increased release of free fatty acid from adipose tissue during fasting. Presumably in adipose tissue the free cholesterol content also tends to increase during such an activation phase. Such a situation may favour the efflux of free cholesterol from adipocytes to possible extracellular acceptor molecules such as high density lipoproteins in plasma.

21. Inhibitors of cholesterol ester hydrolase

Potential inhibitors of cholesterol ester hydrolase have been studied in vivo and in vitro. The isolated rat adrenal cell preparations proved to be a useful system for the study of such inhibitors of cholesterol ester hydrolase and of steroidogenesis. As discussed earlier in this review, mitochondria of adrenal cells contain the cholesterol side chain cleavage enzymes capable of converting cholesterol to pregnenolone, which is considered to be one of the rate-limiting steps in steroidogenesis [93,94]. Mitochondria utilise free or non-esterified cholesterol in this side chain cleavage event and consequently there must be a mechanism for the continuous supply of cholesterol to mitochondria. In the adrenal gland cholesterol is stored by esterification of the sterol to long chain unsaturated fatty acids and the release of free cholesterol through a cholesterol ester hydrolase may be a factor in the overall rate of steroido-genesis. The administration of ACTH to hypophysectomized animals increases the rate of synthesis and secretion of corticosterone and decreases the amount of cholesterol stored as cholesterol ester in the adrenal cells [9,52]. As reviewed in this paper, the cytosol of the adrenal cell contains an activatable cholesterol ester hydrolase.

Valuable co-lateral evidence on the possible significance of the activation of the cholesterol ester hydrolase in the physiology of the adrenal gland has been produced by certain inhibitor studies. Organophosphate insecticides which are known to inhibit lecithin-cholesterol-acyl-tranferase were tested in isolated adrenal cell preparations and found to influence steroidogenesis in these cells and also shown to influence cholesterol ester hydrolase [95]. In rat adrenal cells cholesterol esterification and cholesterol hydrolysis were inhibited by organo-phosphates and it was shown that there was a close correlation between the inhibition of cholesterol ester hydrolase and the inhibition of steroidogenesis. These inhibitors blocked the cytosolic cholesterol ester hydrolase both in the presence and in the absence of exogenous ATP and cyclic AMP. These organophosphate compounds inhibited the phosphorylated enzyme rather than inhibiting the process leading to phosphorylation [95]. In these studies the administration of such an organophosphate compound in the drinking water to rats produced a significant depression of the plasma corticosterone concentration during the rising phase of the normal diurnal rhythm of plasma corticosterone. When such organophosphate treated animals were stressed, the expected fall in the adrenal cholesterol ester concentration did not occur and there was a failure to activate the adrenal cholesterol ester hydrolase enzyme. Such evidence, in the rat, supports the concept that the secretion of corticosterone from the adrenal cortex is intimately associated with cholesterol ester depletion and that the latter is linked to the cholesterol ester hydrolase activity. The employment of this organophosphate inhibitor has emphasised that the release of free cholesterol from esterified cholesterol in the adrenal gland is possibly a significant event in the mechanism of the synthesis and release of steroid hormones by the adrenal. Although the cholesterol utilised in steroidogenesis can be derived from de novo synthesis and by the uptake of plasma cholesterol by the adrenal cells nevertheless the controlled release of free cholesterol from stored esterified cholesterol may well be a significant factor in the acute response in steroidogenesis. It is obvious that in dealing with compounds like the organophosphates which are extremely reactive molecules the precise mode of action in vivo of these compounds is difficult to establish. Nevertheless in these studies using organophosphates on adrenal cells it is possible to show that these insecticides act beyond the site of action of ACTH and beyond the adenyl cyclase event. Since the cholesterol ester hydrolase in the cytosol of the adrenal cells seems quite sensitive to these organophosphates it is reasonable to conclude that cholesterol ester hydrolase enzymes may be a target for these inhibitors. The subsequent failure to produce adequate amounts of corticosterone in such organophosphate treated animals might be due to interference with the activation of the cholesterol ester hydrolase enzyme [95].

22. The effect of dietary lipids on adrenal cortical lipid composition and adrenal cortical cholesterol ester hydrolase activity

The adrenal cortex in most species is rich in esterified cholesterol. Within the adrenal cortical cells, cholesterol ester is present as lipid droplets in the cytoplasm of the cells. The endoplasmic reticulum of adrenal cells has the acetyl coenzyme A-cholesterol-acyl transferase enzyme (ACAT) which effectively esterifies free or non-esterified cholesterol in the presence of CoA, ATP and a long-chain fatty acid. The ACAT enzyme system appears to be much more effective when supplied with monoenoic fatty acids or polyunsaturated fatty acids. It is likely that this is the reason why the cholesterol esters in the cytoplasm of adrenal cells are found to contain cholesterol esterified to oleic acid, linoleic acid, linolenic acid, arachidonic acid and adrenic acid.

Thus, the common non-essential unsaturated fatty acids and the essential fatty acids are used to esterify cholesterol in this organ. The cholesterol esters in the adrenal glands can be rapidly mobilised for production of steroid hormones [9,52] and as the obligatory starting material for steroid hormone production is non-esterified or free cholesterol, then the controlled hydrolysis of esterified cholesterol is an important step in the release of cholesterol for steroidogenesis. Attention has been drawn to the evidence in support of the concept that one role of ACTH in the adrenal cortex is to activate an enzymic cascade resulting in the controlled hydrolysis of adrenal cortical sterol esters.

In most mammals the type of fatty acid laid down in various tissues depends upon a variety of factors, however, and the dietary intake of triglyceride influences the qualitative and quantitative aspects of the fatty acids esterified to glycerol and cholesterol in mammalian tissues. Thus, if animals are fed a diet low in fat, then the fatty acids which are deposited in various tissues, including the adrenal gland, tend to be oleic acid and certain other unsaturated fatty acids of the non-essential type. On the other hand when the diet of animals is supplemented with liberal amounts of linoleic acid then the cholesterol esters, triglycerides and phospholipid in the cells of such animals reflect the dietary lipid composition and are accordingly high in linoleate and arachidonate esters. When rats are maintained on an essential fatty acid (EFA) deficient diet the adrenal cortical cholesterol ester composition changes so that the predominant esters are C_{22}, $\Delta 7$, 10, 13, C_{20}, $\Delta 5$, 8, 11 and oleate, $-C_{18}$, $\Delta 9$ [96].

One of the consequences of essential fatty acid deficiency in its more severe state is a metabolic abnormality which may in part be attributed to adrenal insufficiency. For this reason considerable attention has been paid to changes which may occur in the adrenal cortex in animals mildly deficient or severely deficient in essential fatty acids. It has been shown that rats maintained for as long as 16 weeks on an EFA deficient diet respond by altering the chemical composition of the cholesterol esterified to fatty acids in the adrenal cortex. The diet is deficient in linoleic acid and under these circumstances there is very little cholesterol linoleate or cholesterol arachidonate found in the lipid droplets in these adrenal cells. Although the composition of cholesterol esters in these adrenal cells is drastically altered as a consequence of the essential fatty acid deficient diet the response of the adrenal cells to ACTH as measured by the ability of the cells to produce corticosterone is little impaired in this state. The extent of hydrolysis of the cholesterol esters in the adrenal cells of EFA deficient rats is comparable to that observed in adrenal cells from normal animals. Thus, although the nature of the cholesterol ester stores in the adrenal cortical cells is altered as a consequence of the essential fatty acid deficiency state the activation of cholesterol ester hydrolase appears to be normal resulting in a normal release of free cholesterol for utilisation by the mitochondria in these cells [96].

It is established that diets containing high concentrations of erucic acid which is a C_{22}, $\Delta 13$ acid, produces a marked effect on the lipid composition of various tissues in the rat. One consequence of a high erucate diet is an accumulation of lipid in heart, skeletal muscle and adrenals. Such high erucate diets retard the growth of young animals while the accumulation of lipids in the adrenals is due to a substantial increase in the amount of cholesterol esters in the adrenal cells [56]. These cells appear to store cholesterol erucate selectively. It has been shown that if animals previously fed an erucate containing diet are subjected to ACTH treatment or subjected to a stressful situation, cholesterol ester depletion in the adrenals of the erucate fed animals is much less than the cholesterol ester depletion observed in the adrenals of control animals [56]. Although the erucate fed animals accumulate substantial amounts of esterified cholesterol in their adrenals such animals exposed to a cold stress produce a lower level of plasma corticosterone than is observed in the control animals [97].

Rats fed a high erucate diet, accumulate cholesterol erucate in the adrenal cortical lipid droplets, but fail to produce a decrease in the adrenal cholesterol erucate concentration when such animals are subjected to stressful situations. Ether anaesthesia stress, which elevates the plasma ACTH concentration, produced a 2-fold increase in activity of adrenal cortical cholesterol ester hydrolase in animals fed a control diet [56], but no significant stimulation of adrenal cortical cholesterol ester hydrolase activity occurred in animals fed a high erucate diet. It was shown that the cholesterol erucate was hydrolysed at rates only about one quarter of the rates of hydrolysis of cholesterol oleate by the adrenal cortical cholesterol ester hydrolase enzyme (Figure 12). It has therefore been concluded that the accumulation of cholesterol erucate in the adrenal cortex of animals fed high erucate diet could be due to the reduced ability of cholesterol ester hydrolase to cleave the cholesterol erucate substrate.

This effect of erucate on the adrenal cortex was only observed when the erucate content of the diet represented about 50% by weight of the triglyceride content of the diet. When the triglyceride content of the diet contained only 15% erucate there appeared to be no harmful effect of this diet on the growth of the animals or on the response of the adrenal cortex to ACTH (unpublished work). These studies suggested there might be two effects of erucate in the adrenal cortex, the first being to change the composition of the cholesterol esters producing a pattern with an increased content of cholesterol erucate, the absolute concentration of cholesterol erucate depending upon the amount of erucic acid in the diet. Secondly, there appears to be another possible effect of dietary erucate. At high erucate concentrations in the diet, this fatty acid appears to prove toxic to the organism producing the retardation of growth in the young growing animal, deposition of lipids in skeletal and cardiac muscle, alteration in mitochondrial function in various tissues and a depression of activatable protein kinase in the adrenal cortical cytosol [56].

23. Activatable cholesterol ester hydrolase in tissues other than the adrenal cortex

The presence of an activatable cholesterol ester hydrolase in the cytosol of adrenal cortical cells has stimulated a search for a similar activatable enzyme in other tissues.

From a steroidogenic standpoint the corpus luteum has many characteristics in common with the adrenal cortex. Both tissues contain substantial stores of esterified cholesterol and both tissues upon stimulation by the appropriate trophic hormone produce an increased secretion of steroid hormones associated with a depletion of esterified cholesterol. It was shown [98] that LH administration to the Parlow rat preparation resulted in a decrease in the cholesterol ester composition of the corpus luteum and an increase in progesterone production. In the same study it was shown that LH treatment to these animals 60 min before killing activated cholesterol ester hydrolase activity in the cytosol of corpora lutea. Subsequently [81] again using the Parlow rat preparation it was shown that LH administration to such female rats produced an activation of the cytosolic cholesterol ester hydrolase and an increase in the ratio of free cholesterol to esterified cholesterol in the lipid droplets derived from the corpora lutea [81]. Using bovine corpora lutea it was shown that the cytosol contained cholesterol ester hydrolase activity capable of being activated by cAMP plus ATP [100–102].

However the activation of bovine corpus luteum cytosolic cholesterol ester hydrolase differs in certain respects to the corresponding enzyme in the adrenal cortex. The rate of activation of the bovine corpus luteum cholesterol ester hydrolase is much slower than the rate of activation of the adrenal cortical or the adipose tissue esterases. It is possible that the corpus luteum has a more active phosphoprotein phosphatase than is found in the adrenal cortex, or perhaps it has no phosphoprotein phosphatase inhibitor or a less effective inhibitor.

However the slower activation of the corpus luteum cholesterol ester hydrolase would be consistent with the much slower rate at which progesterone has to be secreted by the corpus luteum.

If the rat is fed excess cholesterol the sterol is sequestered in the liver as cholesterol esters. If rat liver is subjected to a standard cell fractionation procedure and the cholesterol ester hydrolase activity measured in each fraction the microsomal fraction, the lysosomes and the cell cytosol contain most of the hydrolase activity [103]. The cell cytosol contained a cholesterol ester hydrolase enzyme with a pH optimum of about 7. This cytosolic hydrolase was not activated by cAMP plus ATP (unpublished observations). Thus although the liver of the rat stores esterified cholesterol this ester is not mobilised in this tissue by means of activation of a cytosolic cholesterol ester hydrolase.

As discussed in this review and elsewhere there is now a great deal of evidence supporting the concept that the supply of non-esterified cholesterol to mitochondria is a highly significant factor in the control of steroidogenesis in the adrenal cortex of some species. This non-esterified or free cholesterol may be derived by de novo synthesis in the cell, by uptake from the plasma lipoproteins involving exchange of free cholesterol or by the controlled hydrolysis of esterified cholesterol in the cell. In this review attention has concentrated upon the activation of cytosolic cholesterol ester hydrolase. A great deal of support has now accumulated in favour of the hypothesis that in the adrenal cytosol a cholesterol ester hydrolase exists whose activity is modulated by phosphorylation. As shown in Fig. 13 there are many facets of the activation and the deactivation of this enzyme which still require clarification. It is unlikely that an accurate view of this cascade process will emerge until the various enzymes involved in this biological amplifier have been purified. When such purified enzymes and naturally occurring inhibitors of the enzymes are available it may then be possible to predict with greater accuracy some of the actions of ACTH on the adrenal cell and LH on the corpus luteum.

130

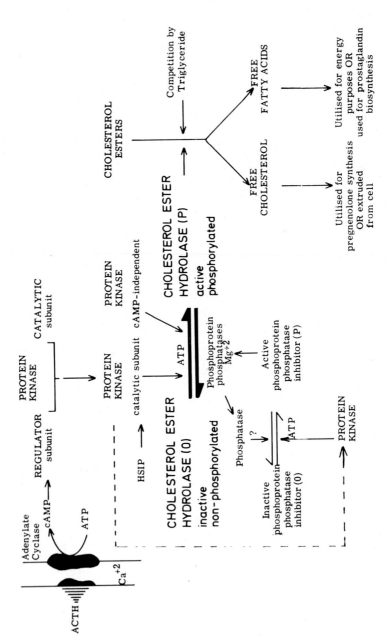

Figure 13. Some of the possible factors which may affect the cytosol cholesterol ester hydrolase in the adrenal cortex. HSIP = high sensitivity inhibitor protein (i.e. the protein kinase inhibitor).

24. Summary

1. The rate limiting event in steroidogenesis in the adrenal cortex in some species may be the delivery of cholesterol to the side chain cleavage reaction centre within the inner cristae of the adrenal mitochondria.

2. The cytosol of the adrenal cell contains a very small supply of free cholesterol but a large reservoir of esterified cholesterol is present as lipid droplets.

3. The cytoplasm of the adrenal cortex contains a cholesterol ester hydrolase which is activatable by phosphorylation. This phosphorylation is effected by a cyclic AMP dependent protein kinase. The protein kinase is kept in a less active state by association with a cyclic AMP binding protein which is the receptor sub-unit of the protein kinase.

4. The presence of cyclic AMP at micromolar concentrations achieves dissociation of the receptor sub-unit from the protein kinase catalytic sub-unit with resultant activation of the protein kinase.

5. There are various other cyclic AMP independent protein kinases found in the adrenal cortex and these enzymes might also play a part in the activation of the enzymes concerned with steroidogenesis.

6. The adrenal cortex contains protein kinase inhibitors. It is possible that some of these inhibitors may modulate the adrenal cortical events associated with the response of the cells to stimulation with ACTH.

7. The cytosol of the adrenal cortex contains a phosphoprotein phosphatase which is able to dephosphorylate the phosphorylated, active, cholesterol ester hydrolase to an inactive, dephosphorylated form.

8. The adrenal cortex contains in the cytoplasm a small molecular weight inhibitor of phosphoprotein phosphatase. This inhibitor exists in two forms, a phosphorylated (active) form and a dephosphorylated (inactive form). The conversion of the inactive to the active inhibitor of phosphoprotein phosphatase is under the control of cyclic AMP-dependent protein kinase.

9. This activatable cholesterol ester hydrolase found in the adrenal cortex may be identical to the activatable triglyceride hydrolase enzyme found in the same cell fraction.

10. Results of inhibitor studies employing potent organophosphates suggest that the rate of hydrolysis of cholesterol esters may be of significance in the acute steroidogenic response of adrenal cells to activation by ACTH.

11. The activation of protein kinase with the subsequent enzymic cascade leading to the controlled hydrolysis of cholesterol esters may not be the exclusive mode of operation of cyclic AMP in adrenal cells.

132

12. The cytosol of rat liver does not appear to contain a cholesterol ester hydrolase activatable by phosphorylation.

References

1. Schulster, D., Burstein, S. and Cooke, B.A. (1976) Molecular Endocrinology of the Steroid Hormones, John Wiley, London.
2. Makin, H.L.J. (ed.), (1975) Biochemistry of the Steroid Hormone, Blackwell, London.
3. Finkelstein, M. and Shaefer, J.M. (1979) Physiol. Rev., 59, 353–406.
4. McKerns, K.W. (ed.) (1968) Functions of the adrenal cortex, Volumes 1 and 2, Appleton-Century-Crofts, New York.
5. Schulster, D. (1974) Adv. Steroid. Biochem. Pharmacol., 4, 233–295.
6. Halkerston, J.D.K. (1975) Adv. Cyclic Nucleotide Res., 6, 99–136.
7. Saffran, M., Grad, B. and Bayliss, M.J. (1952) Endocrinology, 50, 639–643.
8. Haynes, R.C., Jr. and Berthet, L. (1975) J. Biol. Chem., 225, 115–124.
9. Garren, L., Ney, R.L. and Davis, W.W. (1965) Proc. Natl. Acad. Sci. USA, 53, 1443–1450.
10. Grahame-Smith, D.G., Butcher, R.W., Ney, R.I. and Sutherland, E.W. (1967) J. Biol. Chem., 242, 5535–5541.
11. Schulster, D., Tait, S.A.S., Tait, J.F. and Mortek, J. (1970) Endocrinology, 86, 487–502.
12. Seeling, S. and Sayers, G. (1973) Arch. Biochem. Biophys., 154, 230–239.
13. Civen, M. and Brown, C.B. (1974) Pestic. Biochem. Physiol., 4, 254–259.
14. Sharma, R.K., Ahmed, N.K. and Shanker, G. (1976) Eur. J. Biochem., 70, 427–433.
15. Rubin, R.P., Laychock, S.G. and End, D.W. (1977) Biochim. Biophys. Acta, 496, 329–338.
16. Laychock, S.G. and Hardman, J.G. (1978) J. Cyclic Nucleotide Res., 5, 335–344.
17. McKenna, T.J., Island, D.P., Nicolson, W.E. and Liddle, G.W. (1978) Steroids, 32, 127–136.
18. Li, C.H. (1962) Rec. Progr. Horm. Res., 18, 1–32.
19. Schwyzer, R. (1977) Ann. N.Y. Acad. Sci., 297, 3–26.
20. Schwyzer, R., Schiller, P., Seelig, S. and Sayers, G. (1971) FEBS Lett., 19, 229–231.
21. Seelig, S., Kumar, S. and Sayers, G. (1972) Proc. Soc. Exp. Biol. Med., 139, 1217–1219.
22. Lowry, P.J. and McMartin, C. (1972) J. Endocrinol., 55, xxxiii.
23. Richardson, M.C. and Schulster, D. (1972) J. Endocrinol., 55, 140–148.
24. Schimmer, B.P., Ueda, K. and Sato, G.H. (1968) Biochem. Biophys. Res. Commun., 32, 806–810.
25. Lefkowitz, R.J., Roth, J., Pricer, W. and Pastan, I. (1970) Proc. Natl. Acad. Sci. USA, 65, 745–752.
26. Slinger, R.C.L. and Civen, M. (1971) Biochem. Biophys. Res. Commun., 43, 993–999.
27. Golder, M.P. and Boyns, A.R. (1972) Biochem. J., 129, 7.
28. Rubin, R.P. and Laychock, S.G. (1978) Ann. N.Y. Acad. Sci., 307, 377–390.
29. Bonnafous, J.C., Fauchiere, J.L., Schlegel, W. and Schwyzer, R. (1977) FEBS Lett., 78, 247–250.
30. Schlegel, W. and Schwyzer, R. (1977) Eur. J. Biochem., 72, 415–424.
31. Tell, G.P., Cathiard, A.M. and Saez, J.M. (1978) Cancer Res., 38, 955–959.
32. Ontjes, D.A, Ways, D.K. and Mahafee, D.D. (1977) Ann. N.Y. Acad. Sci., 297, 295–313.
33. Latner, A.L. Cook, D.E. and Solanki, K.U. (1977) Biochem. J., 164, 477–480.
34. Beall, R.J. and Sayers, G. (1972) Arch. Biochem. Biophys., 148, 70–76.
35. Haksar, A., Mandsley, D.V., Kimmel, G.L. and Peron, F.G. (1974) Biochim. Biophys. Acta, 362, 356–365.
36. Mackie, C., Richardson, M.C. and Schulster, D. (1972) FEBS Lett., 23, 345–348.

37. Saez, J.M., Evain, D. and Gallet, D. (1978) J. Cyclic Nucleotide Res., 4, 311–321.
38. Podesta, E.J., Milani, A., Steffen, H. and Neher, R. (1979) Biochem. J., 180, 355–363.
39. Haynes, R.C., Koritz, S.B. and Peron, F.G. (1959) J. Biol. Chem., 1421–1423.
40. Gill, G.N. and Garren, L.D. (1969) Proc. Natl. Acad. Sci. USA, 63, 512–519.
41. Gill, G.N. and Garren, L.D. (1971) Proc. Natl. Acad. Sci. USA, 68, 786–790.
42. Gill, G.N. and Garren, L.D. (1970) Biochem. Biophys. Res. Commun., 39, 335–343.
43. Walton, G.M. and Garren, L.D. (1970) Biochemistry, 9, 4223–4229.
44. Garren, L.D., Gill, G.N., Masui, H. and Walton, G.M. (1971) Rec. Prog. Horm. Res., 27, 433–478.
45. Cochet, C., Job, D. and Chambaz, E.M. (1977) FEBS Lett., 83, 53–58.
46. Job, D., Cochet, C., Pirollet, F. and Chambaz, E.M. (1979) FEBS Lett., 98, 303–308.
47. Nimmo, H.G. and Cohen, P. (1977) Adv. Cyclic Nucleotide Res., 8, 145–266.
48. Cochet, C., Job, D. and Chambaz, E.M. (1977) FEBS Lett., 83, 59–62.
49. Sayers, G., Sayers, M.A., Fry, E.G., White, A. and Long, C.N.H. (1944) Yale J. Biol. Med., 16, 361–392.
50. Sayers, G.M., Sayers, M.A., Liang, T. and Long, C.N.H. (1946) Endocrinology, 38, 1–9.
51. Davis, W.W. and Garren, L.D. (1966) Biochem. Biophys. Res. Commun., 24, 805–810.
52. Trzeciak, W.H. and Boyd, G.S. (1973) Eur. J. Biochem., 37, 327–333.
53. Shima, S., Mitsunaga, M. and Nakao, T. (1972) Endocrinology, 90, 808–814.
54. Matsayama, H., Ruhmann-Wennhold, A. and Nelson, D.H. (1971) Endocrinology, 88, 692–695.
55. Cook, D.M., Greer, M.A. and Kendall, J.W. (1972) Proc. Soc. Exp. Biol. Med., 139, 972–974.
56. Beckett, G.J. and Boyd, G.S. (1975) Eur. J. Biochem., 53, 335–342.
57. Boyd, G.S. and Trzeciak, W.H. (1973) Ann. N.Y. Acad. Sci., 212, 361–377.
58. Chen, L. and Morin, L. (1971) Biochim. Biophys. Acta, 231, 194–197.
59. Pittman, R.C., Khoo, J.C. and Steinberg, D. (1975) J. Biol. Chem., 250, 4505–4511.
60. Pittman, R.C. and Steinberg, D. (1977) Biochim. Biophys. Acta, 487, 431–444.
61. Naghshineh, S., Treadwell, C.R., Gallo, L. and Vahouny, G.V. (1974) Biochem. Biophys. Res. Commun., 61, 1076–1082.
62. Trzeciak, W.H. and Boyd, G.S. (1974) Eur. J. Biochem., 46, 201–207.
63. Beckett, G.J. and Boyd, G.S. (1975) Biochem. Soc. Trans., 3, 949–950.
64. Wallat, S. and Kunau, W.-H. (1976) Hoppe-Seyler's Z. Physiol. Chem., 357, 949–960.
65. Beckett, G.J. and Boyd, G.S. (1977) Eur. J. Biochem., 72, 223–233.
66. Naghshineh, S., Treadwell, C.R., Gallo, L.L. and Vahouny, G.U. (1978) J. Lipid Res., 19, 561–569.
67. Khoo, J.C., Steinberg, D., Huang, J.J. and Vagelos, P.R. (1976) J. Biol. Chem., 251, 2882–2890.
68. Khoo, J.C., Sperry, P.J., Gill, G.N. and Steinberg, D. (1977) Proc. Natl. Acad. Sci. USA, 74, 4843–4847.
69. Gorban, A.M.S. and Boyd, G.S. (1977) FEBS Lett., 79, 54–58.
70. Huttunen, J.K., Ellinberg, J., Pittman, R.C. and Steinberg, D. (1970) Biochim. Biophys. Acta, 218, 333–346.
71. Huttunen, J.K. and Steinberg, D. (1971) Biochim. Biophys. Acta, 239, 411–427.
72. Heller, R.A. and Steinberg, D. (1972) Biochim. Biophys. Acta, 270, 65–73.
73. Tsai, S.-C., Fales, H.M. and Vaughan, M. (1973) J. Biol Chem., 248, 5278–5281.
74. Khoo, J.C. and Steinberg, D. (1975) Methods Enzymol., 35, 181–189.
75. Severson, D.L., Khoo, J.C. and Steinberg, D. (1977) J. Biol. Chem., 252, 1484–1489.
76. Steinberg, D. (1976) Adv. Cyclic Nucleotide Res., 7, 157–198.
77. Hülsmann, W.C., Geelhoed-Mieras, M.M., Jansen, H. and Houtsmuller, U.M.T. (1979) Biochim. Biophys. Acta., 572, 183–187.
78. De Lange, R.T., Kemp, R.G., Riley, W.D., Cooper, R.A. and Krebs, E.G. (1968) J. Biol. Chem., 243, 2200–2215.
79. Corbin, J.D., Reimann, E.M., Walsh, D.A. and Krebs, E.G. (1970) J. Biol. Chem., 245, 4849–4851.

134

80. Cohen, P. (1978) Curr. Top. Cell. Regul., 14, 118–196.
81. Beckett, G.J. (1975) Ph.D. Thesis, Edinburgh University.
82. Merlevede, W. and Riley, G.A. (1966) J. Biol. Chem., 241, 3517–3524.
83. Huang, F.L., Tao, S.-H. and Glinsmann, W.H. (1977) Biochem. Biophys. Res. Commun., 78, 615–623.
84. Kalala, L., Goris, J. and Merlevede, W. (1977) Hoppe-Seyler's Z. Physiol. Chem., 358, 575–581.
85. Berridge, M.J. (1975) Adv. Cyclic Nucleotide Res., 6, 1–98.
86. Berridge, M.J. and Rapp, P. (1977) Mechanisms of Action. (Cramer, H. & Schultz, J., eds.), Wiley, London and New York.
87. Sayers, G., Beall, R.J. and Seelig, S. (1972) Science, 175, 1131–1133.
88. Farese, R.U. and Prodente, W.J. (1977) Biochim. Biophys. Acta, 497, 386–395.
89. Mason, J.I., Arthur, J.R. and Boyd, G.S. (1978) Mol. Cell. Endocrinol., 10, 209–223.
90. Cohen, P., Burchell, A., Foulkes, J.G. and Cohen, P.T.W. (1978) FEBS Lett., 92, 287–293.
91. Wang, J.H. (1977) Mechanisms of Action (Cramer, H. & Schultz, J., eds.), Wiley, London and New York.
92. Lundberg, B., Klemets, R. and Löugren, T. (1979) Biochim. Biophys. Acta, 572, 492–501.
93. Harding, B.W., Bell, J.J., Oldham, S.B. and Wilson, L.D. (1968) in: Functions of the Adrenal Cortex (McKerns, K.W., ed.), Vol. 2, pp. 831–896, Appleton-Century-Crofts, New York.
94. Koritz, S.B. (1968) in: Functions of the Adrenal Cortex (McKerns, K.W., ed.), Vol. 1, pp. 27–48, Appleton-Century-Crofts, New York.
95. Civen, M., Brown, C.B. and Morris, R.J. (1977) Biochem. Pharmacol. 26, 1901–1907.
96. Vahouny, G.V., Hodges, V.A. and Treadwell, C.R. (1979) J. Lipid Res., 20, 154–161.
97. Walker, B.L. and Carney, J.A. (1971) Lipids, 6, 797–804.
98. Behrman, H.R. and Armstrong, D.T. (1969) Endocrinology, 85, 474–480.
99. Flint, A.P.F., Grinwich, D.L. and Armstrong, D.T. (1973) Biochem. J., 132, 313–321.
100. Goldstein, S. and Marsh, J.M. (1973) in: Protein Phosphorylation in Control Mechanisms (Huijing, F. and Lee, E.Y.C., eds.), Miami Winter Symposia, Vol. 5, Academic Press, New York and London.
101. Marsh, J.M. (1975) Adv. Cyclic Nucleotide Res., 6, 137–200.
102. Bisgaier, C.L., Treadwell, C.R. and Vahouny, G.V. (1979) Lipids, 14, 1–4.
103. Deykin, D. and Goodman, D.W.S. (1962) J. Biol. Chem., 237, 3649–3656.

CHAPTER 6

The hormonal control of triacylglycerol synthesis

HUGH G. NIMMO

1. Introduction

Fatty acids, stored as triacylglycerol in adipose tissue and liver, constitute the major energy reserve of most higher animals. It is therefore not surprising to find that the synthesis and degradation of triacylglycerols is under hormonal control. In this article I will attempt to summarise our current knowledge of mechanisms which may be involved in the hormonal control of triacylglycerol synthesis in adipose tissue and liver. Considerable progress in this area has been made in the last three years, and the experimental approaches used have largely been based on our detailed knowledge of mechanisms involved in the control of glycogen metabolism and triacylglycerol breakdown. The control of glycogen metabolism has been reviewed extensively (Chapter 1 and reference 1) but I will outline briefly what is known of the control of triacylglycerol breakdown (lipolysis) in adipose tissue to allow comparison with triacylglycerol synthesis.

1.1. Mechanisms involved in the control of lipolysis

The rate-limiting enzyme in adipose tissue lipolysis, hormone-sensitive lipase (HSL), can be phosphorylated and activated by cyclic AMP dependent protein kinase (see [2] for a review). HSL has never been purified to homogeneity and thus its phosphorylation has not been studied in detail. However, there is good indirect evidence that phosphorylation of the enzyme does occur in intact tissue incubated with hormones. For example, the degree of activation of HSL caused by cyclic AMP dependent protein kinase is considerably less in extracts prepared from hormone-treated tissue than in extracts prepared from control tissue [2,3]. The maximal degree of activation of HSL

Cohen (ed.) Recently discovered systems of enzyme regulation by reversible phosphorylation
© Elsevier/North-Holland Biomedical Press, 1980

that can be caused by phosphorylation was reported to be rather low and variable (50–100% for rat adipose tissue HSL, up to 1000% for chicken adipose tissue HSL [2]). Since hormones can increase the rate of lipolysis in rat adipose tissue by much greater factors [4] this might suggest that mechanisms in addition to the phosphorylation of HSL are involved in the stimulation of lipolysis. An alternative explanation is that the assay conditions used for HSL do not accurately reflect the in vivo state: the activation observed under in vivo conditions might be much greater than that observed in vitro. There is some evidence for this: for example, Khoo et al. [5] have recently shown that the degree of activation of chicken adipose tissue diacylglycerol lipase (which may be identical to HSL [6]) by cyclic AMP dependent protein kinase is very dependent on the ionic strength used in the assay.

On the other hand, Wise and Jungas [7] have postulated the existence of an additional mechanism for the hormonal activation of lipolysis. They showed that treatment of adipose tissue with adrenalin caused an increase in the rate of degradation of *endogenous* triacylglycerol in a homogenate prepared subsequently: treatment of a control homogenate with cyclic AMP dependent protein kinase increased the rate of hydrolysis of *exogenous* triacylglycerol. Thus, while cyclic AMP mediated phosphorylation of HSL does seem to be involved in the hormonal control of lipolysis, other mechanisms may also be involved.

1.2. The enzymes involved in triacylglycerol synthesis

Triacylglycerols may be synthesised in adipose tissue via the esterification of *sn*-glycerol 3-phosphate, dihydroxyacetone phosphate or mono-acylglycerol. Glycerol phosphate seems to be quantitatively the most important precursor [8,9]. The pathways of triacylglycerol synthesis from glycerol phosphate and dihydroxyacetone phosphate are outlined in Figure 1. There is some disagreement as to whether the acylations of glycerol phosphate and dihydroxyacetone phosphate are catalysed by a single enzyme. Schlossman and Bell [8] studied the acylation system in rat adipose tissue microsomes and concluded that glycerol phosphate acyltransferase (GPAT) was identical to dihydroxyacetone phosphate acyltransferase (DHAPAT): the two activities responded similarly to *N*-ethylmaleimide, trypsin and detergent, and each substrate was a competitive inhibitor of the acylation of the other. On the other hand Dodds et al. [10] found no correlation between the adipose tissue GPAT and DHAPAT activities of rats fed on different diets and concluded that different enzymes were involved.

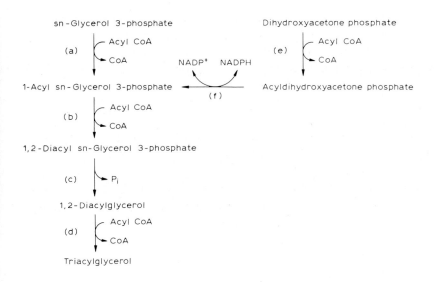

Figure 1. Pathways of triacylglycerol synthesis in adipose tissue. The enzymes involved are (a) *sn*-glycerol 3-phosphate 1-acyltransferase (GPAT), (b) 1-acylglycerol 3-phosphate acyltransferase, (c) phosphatidate phosphatase (d) diacylglycerol acyltransferase, (e) dihydroxyacetone phosphate acyltransferase (f) 1-acylglycerol 3-phosphate dehydrogenase.

In adipose tissue the enzymes fatty acyl CoA synthase, GPAT and diacyl-glycerol acyltransferase are each almost exclusively located in the microsomes [8,11,12]. In contrast, in liver fatty acyl CoA synthase and GPAT are found in both mitochondria and microsomes [13–15]. It has been shown that these two GPAT activities represent distinct isoenzymes which differ in their kinetic properties, heat stabilities and sensitivities to thiol group reagents [15]. The rat adipose tissue microsomal GPAT is very similar to the liver microsomal enzyme in its response to heat treatment and thiol group reagents ([8,16] and H.G. Nimmo, unpublished results), and may indeed be identical to it.

Very little is known about phosphatidate phosphatase in adipose tissue. In liver, phosphatidate phosphatase activity can be detected both in the microsomes and in the soluble fraction (see, e.g. [17]). It is not yet clear whether two distinct enzymes are involved. Sturton et al. [17] reported that the phosphatidate phosphatase activities of the two fractions change similarly in response to various dietary changes, and they suggested that a single enzyme was responsible for the activities in the two fractions. Hubscher et al. [18] showed that maximal rates of glycerolipid synthesis are normally obtained only when the soluble fraction is added back to the microsomal fraction, so it appears that both activities may be involved in glycerolipid synthesis in vivo. It is not yet clear whether the phosphatidate phosphatase involved in triacylglycerol synthesis in adipose tissue is microsomal or soluble. None of the enzymes involved in triacylglycerol synthesis in adipose tissue has yet been purified and in general very little is known about their properties or control.

2. Effects of hormones on triacylglycerol synthesis in adipose tissue

The hypothesis that triacylglycerol metabolism is controlled in a manner analogous to the control of glycogen metabolism in skeletal muscle is very attractive. It leads to the prediction that lipolytic hormones should inhibit esterification. In fact it is well known that adrenalin, glucagon, ACTH and TSH actually stimulate esterification as well as lipolysis in rat fad pads (see, e.g. [4]). This paradoxical result may be caused by the dramatic increase in the availability of free fatty acids, the precursors for triacylglycerol synthesis, that occurs in response to lipolytic hormones. It is not impossible that such hormones should decrease the activities of some or all of the enzymes of the esterification pathway (assayed in vitro) although the flux through the pathway is actually increased.

There is now considerable evidence that insulin can increase the rate of triacylglycerol synthesis in adipose tissue, an effect that is not secondary to its effects on glucose uptake and fatty acid synthesis [19,20]. For example, insulin redirects the metabolism of fructose by adipocytes to favour triacylglycerol synthesis and this effect is considerably greater than its effect on fructose uptake [20]. Thus the esterification pathway is subject to hormonal control. However the rate-limiting step in the pathway has not been identified, largely because the enzymes and intermediates are membrane-bound so that conventional 'crossover analysis' has not been possible. A number of workers have therefore adopted a different approach: they have studied the effects of adding hormones to fat cells on enzyme activity in tissue fractions prepared subsequently. Any change in enzyme activity detected in this way must persist during the preparation of the tissue extract or subcellular fraction and may be caused by a covalent modification. So far significant effects on four of the five enzymes involved in triacylglycerol synthesis have been reported.

2.1. Fatty acyl CoA synthase

Jason et al. [11] incubated rat adipocytes in the presence or absence of insulin:

they then isolated microsomes and found that insulin caused a 2-fold increase in the specific activity of the fatty acyl CoA synthase. The effect was rapid (complete in 2 min) and was maximal at 80–100 μunits/ml insulin. Insulin caused a 2-fold increase in the V_{max} and the K_m for CoASH but did not affect the K_m values for ATP and oleic acid. Sooranna and Saggerson [21] assayed fatty acyl CoA synthase in homogenates prepared after freeze-stopping adipocyte incubations. They reported that 0.63 μM adrenalin caused a 30–40% decrease in the activity of the synthase but they did not rule out the possibility that this effect was secondary to an increase in the rate of lipolysis. A time-course showed that the effect of adrenalin was relatively slow: maximum inhibition was observed only after incubation of the cells with hormone for 30 min. The effects of adrenalin were blocked by insulin and by the β-blocker propanolol, but insulin alone (at 20 munits/ml) had no effect on the activity of fatty acyl CoA synthase. Evans and Denton [22] also found no effect of insulin on the activity of this enzyme. No group has yet reported the effects of a hormone on fatty acyl CoA synthase to be reversible and the mechanism(s) responsible for these hormonal effects has yet to be elucidated.

2.2. Glycerol phosphate acyltransferase

The effects of adrenalin and insulin on the activity of GPAT have been extensively studied by Sooranna and Saggerson [23–26]. In these experiments incubations of adipocytes were freeze-stopped and GPAT was assayed in homogenates prepared from the frozen tissue. Adrenalin (0.63 μM) caused up to 50% inactivation of GPAT in adipocytes incubated without a carbohydrate source [23]. It is not yet known whether adrenalin affects the V_{max} of the enzyme or its K_m values for its substrates. The effect did not seem to be secondary to a stimulation of lipolysis because incubation of the adipocytes in the presence of added palmitate did not affect GPAT [24]. The effect of adrenalin on GPAT was relatively slow: it was not significant before 45 min incubation (for cells in the absence of carbohydrate) or 30 min (for cells in the presence of 5 mM glucose) [25]. No reversal of the inactivation caused by adrenalin has yet been detected, even after incubation of the tissue extract for 70 min at 30°C [25]. However, no attempts to reverse the effect by adding protein phosphatases or by incubating the extract under conditions likely to promote dephosphorylation have yet been made.

The effects of adrenalin on GPAT have also been observed using fat cells from starved animals and when dihydroxyacetone phosphate was used as an acyl acceptor instead of glycerol phosphate [26], supporting the idea that a single enzyme may be responsible for the acylation of both substrates [8]. The inactivation caused by adrenalin was abolished by insulin (20 munits/ml) or by propanolol, but the effects of insulin alone depended on whether or not a carbohydrate source was present [23]. In the absence of carbohydrate insulin caused a slight stimulation of GPAT; in the presence of 5 mM glucose it had no significant effect, and in the presence of 5 mM fructose it actually caused slight inactivation of GPAT [23]. The physiological significance of these effects of insulin on GPAT are questionable. The effects of adrenalin are more easy to rationalise and a possible mechanism for these effects is suggested by the observation (see below) that cyclic AMP dependent protein kinase can cause inactivation of GPAT.

2.3. Phosphatidate phosphatase

Cheng and Saggerson [27,28] studied the effects of noradrenalin and insulin on phosphatidate phosphatase using a procedure similar to that used for the studies on GPAT [23] (see above). They found that noradrenalin caused a time-dependent decrease in the Mg^{2+}-dependent phosphatidate phosphatase activity but that it had no effect on the activity observed in the absence of Mg^{2+} ions (which was about 10% of the total activity) [27]. The inactivation reached 65% after 60 min incubation with 0.5–2.0 μM noradrenalin. They found that insulin (0.2 munits/ml) and propanolol could block the inactivation and also that insulin could cause reactivation of the enzyme in cells which had previously been exposed to noradrenalin [28]. Incubation of adipocytes in the presence of palmitate caused slight inactivation of phosphatidate phosphatase but the authors concluded that the observed effects of hormones could not be attributed simply to changes in the levels of fatty acids in the incubation medium [28]. However no mechanism has yet been proposed to account for these hormonal effects on the activity of phosphatidate phosphatase.

2.4. Diacylglycerol acyltransferase

Sooranna and Saggerson [29] reported that incubation of rat adipocytes with adrenalin caused an inactivation of diacylglycerol acyltransferase: the protocol used was similar to that employed in the studies of GPAT (see above). Their assay system was based on the incorporation of labelled palmitoyl CoA into triacylglycerol. They observed some incorporation (about 40% of the total) in the absence of added diacylglycerol. This activity presumably reflects the level of endogenous acceptors in the homogenate and it was not affected by hormone. However, the activity observed in the presence of added diacylglycerol was affected: incubation of the fat cells with 1.0 μM adrenalin for 60 min caused a 50% loss of the diacylglycerol-dependent activity. Insulin (2 munits/ml) blocked this effect of adrenalin but had no effect on the activity of the enzyme in the absence of adrenalin. No mechanism for this effect of adrenalin on the activity of diacylglycerol acyltransferase has yet been proposed and the possibility that it is secondary to a stimulation of lipolysis and an increased level of non-esterified fatty acids has not been conclusively ruled out.

2.5. Summary

These results have stimulated considerable interest in the mechanisms involved in the control of triacylglycerol synthesis but a number of reservations about them remain. Firstly, the effects of adrenalin and noradrenalin discussed above occur relatively slowly and indeed may not be physiologically significant. Secondly, there is no evidence to suggest that the effects can be reversed in vitro, as would be expected for a covalent modification such as phosphorylation. Thirdly, it is still possible that the effects of the catecholamines on fatty acyl CoA synthase, phosphatidate phosphatase and diacylglycerol acyltransferase result from inactivation caused by elevated levels of non-esterified fatty acids. Nevertheless the results are clearly sufficiently interesting as to warrant further studies on the regulatory properties of these enzymes. The results of some studies on GPAT are presented below.

3. Studies of the regulation of glycerol phosphate acyltransferase in adipose tissue

The fact that adrenalin could cause inactivation of rat adipose tissue GPAT suggested that cyclic AMP dependent protein kinase might be involved in the control of GPAT. It was therefore decided to test the possibility that GPAT is controlled by a phosphorylation–dephosphorylation mechanism. GPAT has never been purified from adipose tissue: indeed, although one procedure for the solubilisation of the rat liver microsomal enzyme has been reported [30] the method has not proved successful in my hands (H.G. Nimmo, unpublished results). We have therefore carried out preliminary experiments with the GPAT activity of freshly isolated microsomes. When rat adipose tissue microsomes were incubated with cyclic AMP dependent protein kinase, cyclic AMP, ATP and MgCl$_2$ the activity of GPAT decreased in a time-dependent manner until only 10–20% activity remained (Figure 2) [31]. Further addi-

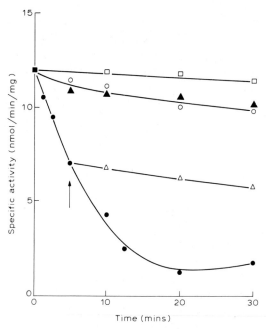

Figure 2. Inactivation of GPAT. Microsomes were prepared from rat epididymal fat pads as described previously [31]. The standard incubation mixture (0.2 ml) contained 0.1 ml microsomes, 5 mM ATP, 10 mM MgCl₂, 10 μM cyclic AMP, 0.05 mg/ml cyclic AMP dependent protein kinase partially purified as described earlier [32]. GPAT activity was assayed as described in [31] and results are expressed in terms of specific activities in nmol/min/mg protein. The symbols represent ●, standard conditions as above; □, ATP omitted; ○, inhibitor protein added (0.7 mg/ml); ▲, cyclic AMP dependent protein kinase omitted and inhibitor protein added; △, in the standard incubation conditions, at the time indicated by the arrow an excess of the inhibitor protein was added.

tions of ATP, cyclic AMP or cyclic AMP dependent protein kinase after 30 min caused no extra decrease in activity. The average inhibition of GPAT observed in a series of eight experiments was 86 ± 9% (mean ± S.D.). As a control the activity of the microsomal enzyme NADPH-cytochrome *c* reductase was tested in several experiments: it was found to be completely unaffected by incubation in these conditions [31].

Some further experiments were carried out to identify the components essential for the inactivation of GPAT. No inactivation was observed in the absence of ATP (Figure 2) or MgCl₂, nor if the ATP was replaced by 5 mM GTP, 5 mM ADP or 5 mM AMPPNP (results not shown). The inactivation could largely be prevented by the presence of the inhibitor protein of cyclic AMP dependent protein kinase [33] or by the absence of the protein kinase itself (Figure 2). In an experiment in which an excess of the inhibitor protein was added during the course of an incubation, the decrease in the activity of GPAT was almost totally halted immediately (Figure 2). These results strongly suggest that GPAT can be inactivated in a phosphorylation reaction catalysed directly by cyclic AMP dependent protein kinase. The fact that addition of the inhibitor protein seems to have an immediate effect argues against the existence of an intervening kinase analogous to phosphorylase kinase (reference 2 and Chapter 1). The slight inactivation of GPAT observed in the absence of cyclic AMP dependent protein kinase and the presence of the inhibitor protein might suggest that small amounts of a cyclic AMP independent inactivating factor are present in the microsomes.

To show conclusively that GPAT can be controlled by a phosphorylation mechanism it will be necessary to demonstrate the incorporation of ^{32}P from [γ-^{32}P]ATP into GPAT. As a first step I have investigated the phosphorylation of microsomal proteins in experiments analogous to those shown in Figure 2. Microsomal preparations contain variable amounts of ATPase activity which can cause considerable problems in phosphorylation experiments. This is illustrated in Figure 3 which shows the results of an experiment in which the initial ATP concentration was 1.0 mM. The level of

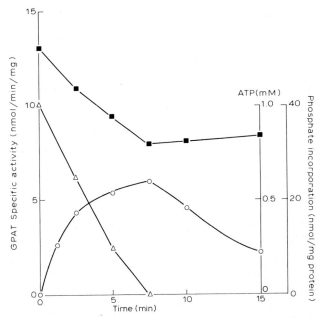

Figure 3. Phosphorylation of rat adipose tissue microsomes. Microsomes were incubated in the standard conditions (Figure 2) except that 1.0 mM [γ-^{32}P]ATP was used. Samples were assayed for GPAT activity (■) [31], ATP concentration (△) [34] and trichloracetic acid–precipitable ^{32}P (○) [35].

ATP remaining in the incubation was monitored and the results show that incorporation of ^{32}P into trichloracetic acid–precipitable material and inactivation of GPAT proceed until all the ATP has been hydrolysed: about 40% inactivation of GPAT had occurred at this stage. After the ATP had been hydrolysed the protein-bound ^{32}P decreased slowly while the GPAT activity remained almost constant. This might imply that GPAT is less rapidly dephosphorylated than other microsomal proteins.

This result is interesting in view of the fact that some early attempts to reactivate GPAT using a purified protein phosphatase were not successful [31]. Reactivation of GPAT was achieved using calf intestinal alkaline phosphatase and this suggested that the inactivation of GPAT caused by phosphorylation was reversible [31]. However, these experiments were open to the criticism that the observed reactivation could have been caused by proteolysis rather than by dephosphorylation. Accordingly a search was made for conditions in which GPAT could be reactivated by a protein phosphatase, and this has now been successful.

Rat adipose tissue microsomes were inactivated as described in the legend to Figure 2. The inactivation was stopped by adding an excess of EDTA and the microsomes were isolated by centrifugation at $100\ 000 \times g$ for 60 min. The microsomes were washed once and finally resuspended in 30 mM triethanolamine hydrochloride–NaOH(pH 7.2) containing 1 mM EDTA, 1 mM dithiothreitol and 50 mM NaCl. The microsomes were then incubated with various concentrations of highly purified protein phosphatase-1 from rabbit skeletal muscle (Chapter 1) in the presence of 5 mM $MgCl_2$ and the GPAT activity was monitored. The results (Figure 4) show that no reactivation was detectable at zero, 0.5 or 1.0 unit/ml phosphatase. Some reactivation was seen using 2.0 units/ml and rapid reactivation was seen using 3.0 units/ml phosphatase.

Protein phosphatase-1 is a multi-functional enzyme which can dephosphorylate phosphorylase a, glycogen synthase b, the β-subunit of phosphorylase kinase and several histones [36,37]. It can be specifically inhibited by two heat-stable protein inhibitors termed inhibitor-1 and inhibitor-2 which may be involved in the regulation of phosphatase activity in vivo [38,39]. It has been suggested that this enzyme is involved in reversing most or all of the phosphorylations catalysed by cyclic AMP dependent protein kinase [37] and the results presented here are compatible with this idea. It should also be noted that a multi-functional protein phosphatase isolated from rat liver is capable of dephosphorylating adipose tissue HSL [40].

It is evident from the results shown in Figure 4 that the rate of reactivation of GPAT is not linearly related to the concentration of the protein phosphatase in the assay. Heat stable phosphatase inhibitor proteins are

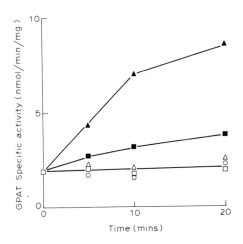

Figure 4. Reactivation of GPAT. Microsomes were inactivated and re isolated as described in the text. They were incubated at 30 °C in the buffer described in the text with the following concentrations of partially purified rabbit muscle protein phosphatase-1 [36] (phosphorylase phosphatase): □, none; ●, 0.5 units/ml; △, 1.0 units/ml; ■, 2.0 units/ml; ▲, 3.0 units/ml. One unit of phosphorylase phosphatase catalyses the release of 1 nmol of P_i from phosphorylase a per min under the standard assay conditions described in [36].

known to occur in adipose tissue [41] and their presence in microsomes could account for these results. I therefore tested the effects of adding rat adipose tissue microsomal protein, either boiled or unboiled, to assays of the protein phosphatase using phosphorylase a as a substrate. The results (Figure 5) show that native microsomes can inhibit the phosphatase. Taken in isolation, this result is difficult to interpret, but it is clear that the supernatant fraction from boiled microsomes can inhibit the phosphatase even more effectively. Preliminary experiments have shown that this inhibitory material is trypsin-labile (H.G. Nimmo, unpublished results). It therefore seems likely that adipose tissue microsomes contain a heat-stable phosphatase inhibitor protein that may be analogous to inhibitor-2 [38]. The heat stable protein, termed inhibitor-1, is only an inhibitor of protein phosphatase-1 after it has been phosphorylated by cyclic AMP dependent protein kinase [39]. It is not yet clear whether the inhibitory activity of the microsomes can be increased by phosphorylation, but the degree of inhibition observed using unphosphory-lated microsomes seems sufficient to account for the lack of effect of the lower concentrations of protein phosphatase shown in Figure 4.

The results presented above show that rat adipose tissue GPAT can be reac-tivated by a highly purified protein phosphatase. This further supports the idea that GPAT can be regulated by a phosphorylation–dephosphorylation

146

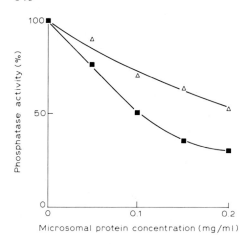

Figure 5. Inhibition of rabbit muscle phosphorylase phosphatase by rat adipose tissue microsomes. Rabbit muscle phosphorylase phosphatase (protein phosphatase-1) was incubated under the standard conditions for assay of phosphatase inhibitors [42] in the presence of the indicated concentration of microsomes (△) or of the supernatant after microsomes had been heated to 100°C for 2 min (■). Control experiments showed that the buffer in which the microsomes were suspended had no inhibitory effect and that the microsomes themselves contained no detectable phosphatase activity.

mechanism. The results shown in Figure 4 are not consistent with the earlier report [31] that protein phosphatase-1 could not reactivate GPAT. There are two possible explanations for this discrepancy. Firstly, different methods have now been used for the isolation of the inactivated GPAT (see above). Secondly, it has become clear that microsomes inhibit the protein phosphatase (Figure 5): this possibility was not fully considered in the earlier work.

It therefore seems clear that rat adipose tissue GPAT can be regulated by a phosphorylation–dephosphorylation mechanism. Ideally one would like to demonstrate the incorporation of ^{32}P into GPAT directly but (since the protein has not yet been purified) this cannot be attempted at the moment. However, much work remains to be done on the physiological significance of the phosphorylation of GPAT. In particular it has not yet been demonstrated that the phosphorylation by cyclic AMP dependent protein kinase is responsible for the effect of adrenalin on the enzyme. There are some differences between the two effects: for example, adrenalin gives only 50% inhibition whereas cyclic AMP dependent protein kinase can cause 80–90% inhibition. This discrepancy may of course be accounted for by differences in the assay conditions used in the two sets of experiments [23,31], or because administration of adrenalin produces only partial phosphorylation of GPAT. If, however, cyclic AMP dependent protein kinase is involved in the action of adrenalin on GPAT, as suggested here, a number of testable predictions can be made. Firstly, the inactivation caused by adrenalin should be reversed by treatment of the homogenate with phosphorylase phosphatase. Secondly, cyclic AMP dependent protein kinase should have little or no further effect on the activity of GPAT in microsomes isolated from tissue previously exposed to adrenalin. This experiment is analogous to one that has already been carried out with HSL [2,3]. We should soon be in a position to assess the physiological significance of the phosphorylation of GPAT much more accurately.

4. Control of lipid synthesis in liver

The esterification of fatty acids in liver is a much more complex pathway than triacylglycerol synthesis in adipose tissue because liver can synthesise several different types of phospholipids as well as triacylglycerols (see, e.g. [43]). Moreover a high proportion of the lipid synthesized by liver is exported in the form of lipoproteins rather than stored. It has already been mentioned that two distinct isoenzymes of GPAT exist in liver, one in the outer mitochondrial membrane and one in the microsomes [14,15], and this may indeed be a reflection of the great complexity of lipid synthesis in this tissue. Fatty acyl CoA synthase activity is also located in both the outer mitochondrial membranes and the microsomes in liver [13], but in neither case has the functional significance of the dual localisation of enzyme activity been clarified.

One of the most interesting features of fatty acid metabolism in liver is that it is subject to dietary control: fed animals preferentially esterify fatty acids whereas fasted animals preferentially oxidise them (see, e.g. [44]). A complete review of this area is outside the scope of this article, but it should be mentioned that McGarry's group has recently suggested that β-oxidation of long chain fatty acids is regulated by inhibition of 'outer' carnitine palmitoyl-transferase by malonyl CoA [45]. This enzyme is thought to be located on the outer surface of the inner mitochondrial membrane [45,46] while the mitochondrial GPAT is located on the inner surface of the outer membrane [47]. The switch between oxidation and esterification of fatty acids may therefore be controlled by a competition between these two enzymes for fatty acyl CoA. This hypothesis is supported by the observations that insulin and diabetes affect the activity of the mitochondrial GPAT much more than that of the microsomal GPAT [48,49].

The synthesis of glycerolipids in rat liver is also under hormonal control. It has been shown that glucagon inhibits the conversion of palmitate to triacylglycerol in hepatocytes but has no effect on the synthesis of phosphatidylcholine and phosphatidylethanolamine [50]. These results suggest that glucagon affects diacylglycerol acyltransferase (which is involved only in triacylglycerol synthesis) rather than GPAT (which is also involved in phospholipid synthesis). However, Heimberg's group have shown that dibutyryl cyclic AMP causes an inhibition of triacylglycerol synthesis in perfused rat liver [51,52]. This was associated with a small but significant inhibition of the GPAT activity of microsomes isolated from the liver after perfusion, whereas the activities of phosphatidate phosphatase and diacylglycerol acyltransferase were actually increased [52]. These results indicate that GPAT becomes inhibited in response to glucagon. Soler-Argilaga et al. [53] have now suggested that Ca^{2+} ions are involved in this response. They showed that in isolated hepatocytes the inhibition of triacylglycerol synthesis by glucagon or dibutyryl cyclic AMP was reduced in calcium-free media and could be mimicked by the ionophore A 23187 provided that Ca^{2+} ions were present. They had previously shown that uptake of Ca^{2+} ions by rat liver microsomes was associated with an inhibition of glycerolipid synthesis [54]. This effect could be observed using 20 μM Ca^{2+} ions provided that ATP was present to allow active uptake. They suggested that Ca^{2+} ions may act as a second messenger in the effects of glucagon on glycerolipid synthesis [53]. These results are very interesting but a number of points require clarification. For example, it has not yet been shown that glucagon actually causes increased uptake of Ca^{2+} ions by liver microsomes. Moreover, the mechanism underlying the effects of dibutyryl cyclic AMP in the perfused liver is not clear. Finally, it is not yet certain that the microsomal GPAT is in fact involved in the regulation of glycerolipid synthesis in liver. Of the two isoenzymes of GPAT in liver, the mitochondrial may be the more important in regulatory terms (see above). There is also considerable evidence suggesting that changes in the activity of phosphatidate phosphatase may be very important in the control of glycerolipid synthesis in liver, particularly in the long term (see, e.g. [17,55]).

The liver microsomal GPAT is very similar or identical to the adipose tissue microsomal GPAT, whereas the liver mitochondrial GPAT is clearly a distinct isoenzyme [15]. Some preliminary experiments have been carried out with the liver enzymes to investigate whether they could also be controlled by phosphorylation and the results are shown in Figure 6. In both cases inactivation was observed. With the microsomal enzyme, only about 40% inactivation was observed, and this was dependent on the presence of added cyclic AMP dependent protein kinase. However, with the mitochondrial

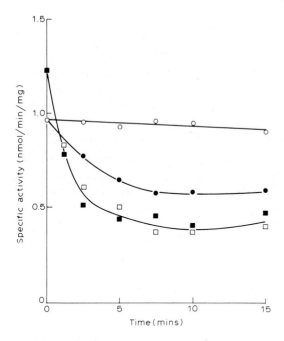

Figure 6. Inactivation of liver mitochondrial and microsomal GPATs. Liver mitochondria (■,□) and microsomes (●,○) were prepared as described in [15]. They were incubated either with 5 mM ATP and 10 mM MgCl₂ (open symbols) or with 5 mM ATP, 10 mM MgCl₂, 10 μM cyclic AMP and 0.05 mg/ml cyclic AMP dependent protein kinase, partially purified as described earlier [32] (closed symbols). Samples were assayed for GPAT activity and the results are expressed as specific activities in nmol/min/mg protein.

GPAT, 50–60% inactivation was observed even in the absence of added protein kinase. These results are only preliminary, and indeed it has not yet been definitely established that phosphorylation is involved (for example, no reactivation studies with phosphatases have yet been attempted). The results may, however, indicate that the mitochondrial GPAT can be regulated by a cyclic AMP independent protein kinase. This suggestion clearly requires further confirmation and much work remains to be done on the physiological roles of the two isoenzymes of GPAT in liver.

5. Summary and conclusions

A number of authors have speculated that triacylglycerol synthesis in adipose tissue may be controlled by phosphorylation mechanisms similar to those involved in the control of glycogen metabolism and of lipolysis. The data

150

presented here suggest that this may indeed be the case. The early studies of lipolysis showed clearly that adrenalin caused an increase in the activity of HSL. The analogous studies of triacylglycerol synthesis have suggested that four of the five enzymes involved may be subject to hormonal regulation. In each case the observed effects are rather small, but this may of course reflect difficulties in designing suitable assay conditions. In only one case, that of GPAT, has cyclic AMP dependent protein kinase been shown to affect enzyme activity in vitro. However, we still know much less about the control of esterification than about the control of lipolysis. In particular the evidence that GPAT can be phosphorylated in vitro is still only indirect and there is as yet no evidence to show that the phosphorylation occurs in intact tissue in response to adrenalin. Further studies of GPAT and the other enzymes involved in esterification are clearly needed, and they should increase our understanding of the organisation and control of membrane-bound metabolic pathways.

Acknowledgements

The work carried out in this laboratory was supported by a grant from the Medical Research Council. Part of Figure 2 is reproduced by permission of the *Biochemical Journal*.

References

1. Nimmo, H.G. and Cohen, P. (1977) Adv. Cyclic Nucleotide Res., 8, 145–266.
2. Steinberg, D. (1976) Adv. Cyclic Nucleotide Res., 7, 157–198.
3. Huttunen, J.K., Heller, R. and Steinberg, D. (1971) Trans. Assoc. Am. Physicians, 84, 162–171.
4. Vaughan, M. and Steinberg, D. (1963) J. Lipid Res., 4, 193–199.
5. Khoo, J.C., Steinberg, D. and Lee, E.Y.C. (1978) Biochem. Biophys. Res. Commun., 80, 418–423.
6. Khoo, J.C., Steinberg, D., Huang, J.J. and Vagelos, P.R. (1976) J. Biol. Chem., 251, 2882–2890.
7. Wise, L.S. and Jungas, R.L. (1978) J. Biol. Chem., 253, 2624–2627.
8. Schlossman, D.M. and Bell, R.M. (1976) J. Biol. Chem., 251, 5738–5744.
9. Dodds, P.F., Gurr, M.I. and Brindley, D.N. (1976) Biochem. J., 160, 693–700.
10. Dodds, P.F., Brindley, D.N. and Gurr, M.I. (1976) Biochem. J., 160, 701–706.
11. Jason, C.J., Polokoff, M.A. and Bell, R.M. (1976) J. Biol. Chem., 251, 1488–1492.
12. Coleman, R. and Bell, R.M. (1976) J. Biol. Chem., 251, 4537–4543.
13. de Jong, J.W. and Hülsmann, W.C. (1970) Biochim. Biophys. Acta 231, 32–47.
14. Daae, L.N.W. and Bremer, J. (1970) Biochim. Biophys. Acta, 210, 92–104.
15. Nimmo, H.G. (1979) Biochem. J., 177, 283–288.
16. Schlossman, D.M. and Bell, R.M. (1977) Arch. Biochem. Biophys., 182, 732–742.

17. Sturton, R.G., Pritchard, P.H., Han, L.Y. and Brindley, D.N. (1978) Biochem. J., 174, 667–670.
18. Hubscher, G., Brindley, D.N., Smith, M.E. and Sedgwick, B. (1967) Nature (London), 216, 449–453.
19. Denton, R.M. and Halperin, M.L. (1967) Biochem. J., 110, 27–38.
20. Sooranna, S.R. and Saggerson, E.D. (1975) Biochem. J., 150, 441–451.
21. Sooranna, S.R. and Saggerson, E.D. (1978) FEBS Lett., 92, 241–244.
22. Evans, G.L. and Denton, R.M. (1977) Biochem. Soc. Trans., 5, 1288–1291.
23. Sooranna, S.R. and Saggerson, E.D. (1976) FEBS Lett., 64, 36–39.
24. Sooranna, S.R. and Saggerson, E.D. (1976) FEBS Lett., 69, 144–148.
25. Sooranna, S.R. and Saggerson, E.D. (1978) FEBS Lett., 90, 141–144.
26. Sooranna, S.R. and Saggerson, E.D. (1979) FEBS Lett., 99, 67–69.
27. Cheng, C.H.K. and Saggerson, E.D. (1978) FEBS Lett., 87, 65–68.
28. Cheng, C.H.K. and Saggerson, E.D. (1978) FEBS Lett., 93, 120–124.
29. Sooranna, S.R. and Saggerson, E.D. (1978) FEBS Lett., 95, 85–87.
30. Yamashita, S. and Numa, S. (1972) Eur. J. Biochem., 31, 565–573.
31. Nimmo, H.G. and Houston, B. (1978) Biochem. J., 176, 607–610.
32. Nimmo, H.G., Proud, C.G. and Cohen, P. (1976) Eur. J. Biochem., 68, 31–44.
33. Walsh, D.A., Ashby, D.C., Gonzalez, C., Calkins, D., Fischer, E.H. and Krebs, E.G. (1971) J. Biol. Chem., 246, 1977–1985.
34. Greengard, P. (1963) in: Methods of Enzymatic Analysis (Bergmeyer, H.U., ed.), pp. 551–558.
35. Walsh, D.A., Perkins, J.P., Brostrom, C.D., Ho, E.S. and Krebs, E.G. (1971) J. Biol. Chem., 246, 1961–1967.
36. Antoniw, J.F., Nimmo, H.G., Yeaman, S.J. and Cohen, P. (1977) Biochem. J., 162, 423–433.
37. Burchell, A., Foulkes, J.G., Cohen, P.T.W., Condon, G.O. and Cohen, P. (1978) FEBS Lett., 92, 68–72.
38. Cohen, P., Nimmo, G.A., Burchell, A. and Antoniw, J.F. (1978) Adv. Enzyme Regul., 16, 97–119.
39. Foulkes, J.G. and Cohen, P. (1979) Eur. J. Biochem., 97, 251–256.
40. Severson, D.K., Khoo, J.C. and Steinberg, D. (1977) J. Biol. Chem., 252, 1484–1489.
41. Severson, D.L. and Sloan, S.K. (1977) Biochem. Biophys. Res. Commun., 79, 1045–1050.
42. Cohen, P., Nimmo, G.A. and Antoniw, J.F. (1977) Biochem. J., 162, 435–444.
43. van Golde, L.M.G. and van den Bergh, S.G. (1977) in: Lipid Metabolism in Mammals (Snyder, F., ed.), Vol. 1, pp. 35–149, Plenum Press, New York.
44. Mayes, P.A. (1970) in: Adipose Tissue (Jeanrenaud, B. and Hepp, D. eds.), pp. 186–195, Academic Press, London.
45. McGarry, J.D., Leatherman, G.F. and Foster, D.W. (1978) J. Biol. Chem., 253, 4128–4136.
46. Hoppel, C.L. and Tomec, R.J. (1972) J. Biol. Chem., 247, 832–841.
47. Nimmo, H.G. (1979) FEBS Lett. 101, 262–264.
48. Bates, E.J., Topping, D.L., Sooranna, S.R., Saggerson, D. and Mayes, P.A. (1977) FEBS Lett., 84, 225–228.
49. Bates, E.J. and Saggerson, D. (1977) FEBS Lett., 84, 229–232.
50. Geelen, M.J.H., Groener, J.E.M., de Haas, C.G.M., Wisserhof, T.A. and van Golde, L.M.G. (1978) FEBS Lett., 90, 57–60.
51. Klausner, H., Soler-Argilaga, C. and Heimberg, M. (1978) Metabolism, 27, 13–25.
52. Soler-Argilaga, C., Russell, R.L. and Heimberg, M. (1978) Arch. Biochem. Biophys., 190, 367–372.
53. Soler-Argilaga, C., Russell, R.L., Werner, H.V. and Heimberg, M. (1978) Biochem. Biophys. Res. Commun., 85, 249–256.

152

54. Soler-Argilaga, C., Russell, R.L. and Heimberg, M. (1977) Biochem. Biophys. Res. Commun., 78, 1053–1059.
55. Brindley, D.N. (1978) in: Regulation of Fatty Acid and Glycerolipid Metabolism (Dils, R. and Knudsen, J., eds.), pp. 31–40, Pergamon Press, Oxford.

CHAPTER 7

Protein phosphorylation in the regulation of muscle contraction

1. Introduction

Contraction in all muscle (and many non-muscle) cells occurs by the same basic mechanism. The contractile apparatus is made up of two sets of filaments which move past each other during contraction, this originally being proposed by Huxley [1] as the sliding filament hypothesis. One set of filaments (the thick filaments) contain as their main protein myosin. Figure 1 shows the essential structural features of the molecule [2]. The rod regions of myosin associate together to form the backbone of the thick filaments, while the head regions project out towards the thin filaments. The head region contains the active site for ATP hydrolysis, and is also the location for the

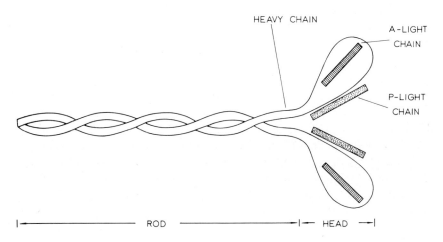

Figure 1. The myosin molecule.

Cohen (ed.) Recently discovered systems of enzyme regulation by reversible phosphorylation
© Elsevier/North-Holland Biomedical Press, 1980

light chains. There are two classes of light chains in all myosins, one being tightly bound (usually called the A-light chain), and the other loosely bound (in vertebrate muscle known as DTNB- or P-light chains) [3]. It is not possible to dissociate the A-light chain without destroying the ATPase activity of the myosin, but the P-light chain will dissociate easily without significant alteration in a number of properties of the myosin [4,5]. Of interest to this article is that in all vertebrate muscle and non-muscle cells, the P-light chain can be phosphorylated by a specific kinase [6–10]. The purpose of this phosphorylation and its control are discussed at some length below (Section 2.2).

The main protein of the thin filament is actin. This exists in solution as an almost spherical monomer of molecular weight 42 000 [11,12], but in the thin filament is polymerised into a double helix with a pitch of 13 actin monomers [13]. The amino acid sequence of actin has been highly conserved throughout evolution [14–16], although there are differences between actins from different muscle and non-muscle types [14,15]. The function of the actin is to interact with the myosin heads of the thick filament during contraction. The myosin head attaches to the actin filament and is thought to undergo a conformational change resulting in a movement of the actin filament relative to the myosin [17].

Control of contraction in all muscles is through the cytoplasmic free Ca^{2+} concentration. In a relaxed muscle the free cytoplasmic Ca^{2+} concentration is less than 10^{-7}M. When the muscle contracts this increases to approximately 10^{-5}M, and the contractile apparatus is activated [18]. However, depending on the type and origin of the muscle both the source of the Ca^{2+}, and the way in which the increase in cytoplasmic Ca^{2+} causes activation of the contractile proteins show considerable differences.

In vertebrate striated muscle, contraction is controlled via the thin filament. This consists, in addition to actin, of two regulatory proteins, tropomyosin and troponin. Tropomyosin is a rod-like protein which runs along the length of the thin filament, each tropomyosin molecule interacting with 7 actins [19]. It has been shown by X-ray diffraction that tropomyosin in relaxed muscle occupies a position on the thin filament that prevents interaction of the myosin heads with actin [20]. In contracting muscle the tropomyosin moves approximately 1.5 nm and allows actin/myosin interaction. This movement is brought about by the binding of Ca^{2+} to troponin. Troponin is located at a fixed point along each tropomyosin molecule, and also interacts with an adjacent actin [21,22]. The ratio of the three major proteins of the thick filament is therefore 7 actin: 1 tropomyosin: 1 troponin. The structure of troponin is discussed in more detail in section B.1 below. When Ca^{2+} binds to troponin there is a conformational change [23] which causes movement of tropomyosin and allows contraction to occur. Of particular relevance to this article is the fact that in certain muscle types troponin can be phosphorylated. This is discussed in Section 2.1.

In smooth muscle and non-muscle cells, regulation of contraction occurs via the myosin in the thick filament [9], and troponin is absent (however, see Ebashi [24–26] for an opposing view). As will be discussed in Section 2.2, this regulation involves Ca^{2+}-stimulated phosphorylation of the myosin P-light chain. When the P-light chain is dephosphorylated the muscle is relaxed, as there is little interaction between the thick and thin filament. An increase in cytoplasmic Ca^{2+} leads to phosphorylation of the P-light chain, and contraction is stimulated. P-light chain phosphorylation is also observed in vertebrate striated muscle [29,30], although its role in the regulation of contraction is at present unclear.

A third type of regulation is seen in molluscan muscle, where the Ca^{2+} binds directly to the myosin head and stimulates the ATPase activity [27,28]. However, as this type of regulation does not involve protein phosphorylation it will not be discussed further in this article.

Just as there are differences between muscle types in the way Ca^{2+} interacts with the contractile proteins, so there are also differences in the source of Ca^{2+} for contraction. Muscle contraction in all cells is initiated by a depolarisation of the cell membrane [18]. In skeletal muscle this leads to release of Ca^{2+} from the sarcoplasmic reticulum, which acts as an internal Ca^{2+} store [31]. Repolarisation of the membrane causes relaxation, which is accompanied by Ca^{2+} uptake into the sarcoplasmic reticulum catalysed by a Ca^{2+}-dependent ATPase [32,33]. In heart muscle, part of the Ca^{2+} arises from the sarcoplasmic reticulum, but a certain amount also crosses the cell membrane from the extracellular medium [34]. Smooth muscle appears to be rather similar to heart, although a higher proportion of the Ca^{2+} may come from the extracellular medium [35]. In heart and smooth muscle, contraction can readily be modified by a number of hormones which change the intracellular concentrations of either cyclic nucleotides or Ca^{2+}. It is of particular interest therefore that proteins present in both sarcoplasmic reticulum and sarcolemma membranes can be phosphorylated by cyclic nucleotide- and Ca^{2+}-stimulated kinases. These phosphorylations often affect the Ca^{2+}-transport ATPases of the membranes, and provide direct mechanisms whereby contraction could be affected by hormones. This is discussed in detail in Section 3.

156

2. Phosphorylation of contractile proteins

2.1. Troponin

Troponin in vertebrate striated muscle is composed of three non-identical subunits [36]. There is a Ca^{2+} binding subunit, troponin-C, an inhibitory subunit, troponin-I, and a tropomyosin-binding subunit, troponin-T. As was described above, Ca^{2+} binding to troponin-C causes a conformational change in the molecule which results in a movement of tropomyosin and initiation of contraction.

The first report of phosphorylation of troponin was that of Bailey and Villar-Palasi [37], who found that incubation of troponin I or whole troponin from skeletal muscle with cyclic AMP-dependent protein kinase resulted in incorporation of phosphate into troponin-I. Further work in the laboratory of Krebs [38,39] showed that with isolated subunits of rabbit skeletal muscle troponin, phosphate could be incorporated into troponin-I and troponin-T by both cyclic AMP-dependent protein kinase and phosphorylase kinase. This phosphate could be removed by phosphorylase phosphatase [40]. However, it became apparent that using intact troponin from skeletal muscle, only very slow rates of phosphorylation were obtained when compared to known physiological substrates for these kinases [39,41,42]. This suggested that the phosphorylation of troponin might be of little physiological significance in skeletal muscle. This has been confirmed by measurement of the phosphate content of troponin + tropomyosin taken from rabbit gracilis muscle before and after treatment with adrenaline or tetanic stimulation [43]. The phosphate content b was unchanged by these treatments, although phosphorylase b to a interconversion had occurred. It appears in skeletal muscle that phosphorylation of troponin does not occur in vivo in response to hormones or contraction. The reason for the phosphorylation seen in vitro with the isolated subunits is probably that denaturation exposes suitable phosphorylation sites normally not available in the native protein. This appears to be a general phenomenon with phosphorylated proteins [44,45], and shows that great care must be taken when studying phosphorylation of proteins in vitro.

Although in the above experiments [43] there was no change in the phosphate content of troponin + tropomyosin, there was 1 mol phosphate/mole complex present throughout. Earlier studies [39,41] had also shown the presence in purified troponin of endogenous phosphate, covalently attached to troponin-T. This phosphate was not readily removed by phospho-protein phosphatase (P.J. England, unpublished observations). It has been assumed that the phosphate present in the troponin + tropomyosin isolated from rabbit muscle is therefore in troponin-T [43]. It appears that this phosphate stays permanently attached to troponin-T and does not have a regulatory role, but may be concerned with the maintenance of structure. A specific troponin-T kinase has been reported [46,47] which is not cyclic AMP-dependent protein kinase or phosphorylase kinase, and may be responsible for the phosphorylation of troponin-T at the time the protein is synthesised.

Early work with isolated troponin from cardiac muscle showed that the intact complex was a good substrate for cyclic AMP-dependent protein kinase, but not phosphorylase kinase [48,49]. This was in contrast to the situation described above with skeletal muscle troponin. The phosphate was incorporated into troponin-I only. The significant difference between cardiac and skeletal muscle troponin-I is that the cardiac subunit has an extra 26 amino acid residues at the N-terminal end [49]. This region has been shown to contain the major phosphorylated site [50], although it has been reported that additional sites can also be phosphorylated.

Phosphorylation of cardiac troponin-I has been demonstrated in vivo in a number of species [51–53]. The experiments were performed with perfused hearts, and showed that in control hearts the phosphate content of troponin-I was 0.3–0.5 mol/mol protein. With adrenaline and other β-agonists there was an increase in phosphorylation to 1–1.5 mol/mol. The time course of the increase in phosphorylation exactly paralleled the increase in force of contraction [51,52], and there was a good correlation between increases in contraction and troponin-I phosphorylation. These results suggested that the phosphorylation of troponin-I could be the cause of the increase in contraction seen with β-agonists. The hypothesis advanced was that phosphorylation of troponin-I would cause an increase in the affinity of troponin for Ca^{2+}, leading to a greater actin/myosin interaction at any given intracellular Ca^{2+} concentration [39,51,52].

However, perfusion of hearts with cardiac glycosides or increased extracellular Ca^{2+} caused an increase in contraction with no increase in troponin-I phosphorylation [54]. It appears therefore that phosphorylation of troponin-I only occurs when the intracellular cyclic AMP concentration is elevated. This also suggests that the enzyme phosphorylating troponin-I in vivo is cyclic AMP-dependent protein kinase [55].

In addition, it was shown using isolated myofibrils that phosphorylation caused a reduction in the affinity of troponin for Ca^{2+}. Heart myofibrils were prepared with troponin-I either in a phosphorylated or a dephosphorylated state [56,57], and the Ca^{2+} sensitivity of the actomyosin ATPase measured. It was found that phosphorylation caused a 5-fold increase in the concentration of Ca^{2+} required for half maximal stimulation of the ATPase. This result, along with the perfusion results with glycosides and increased Ca^{2+}, suggests that the phosphorylation of troponin-I cannot be the mechanism by which increases in contractility are brought about.

Phosphorylation of troponin-I in heart is still important, however. It can be calculated that during the phosphorylation of troponin-I induced by adrenaline, greater than 50% of the cyclic AMP-dependent protein kinase is required to catalyse the reaction [52]. This makes troponin-I a major substrate for protein kinase in heart. One possible explanation for the role of this phosphorylation is in mediating the decrease in relaxation time of the heart seen with adrenaline [58]. Although adrenaline causes changes in rates of Ca^{2+} fluxes in the heart (Section 3.1), a direct effect on the contractile proteins could also be important. A decrease in Ca^{2+} affinity of troponin-I when phosphorylated indicates either a decrease in the rate constant for Ca^{2+} binding (the 'on' rate), or an increase in the rate constant for Ca^{2+} release (the 'off' rate). Studies of many enzymes indicate that changes in 'off' rates occur much more frequently than changes in 'on' rates, and so it is likely that phosphorylation of troponin-I causes an increase in the rate of Ca^{2+} release. Although there is no evidence that release of Ca^{2+} from troponin is the rate-limiting step in the relaxation of cardiac muscle, an increase in this rate on phosphorylation could contribute to the decreased relaxation time. Unfortunately is has not been possible to study the rates of Ca^{2+} binding and release from troponin directly, as phosphorylation of troponin-I has no effect on the Ca^{2+} affinity of purified cardiac troponin [59].

2.2. Myosin light chains

The head region of the myosin molecule contains a loosely bound light chain of molecular weight 19 000 (originally called the DTNB-light chain). As this light chain has been shown to be phosphorylated in all vertebrate systems, it is now commonly called the P-light chain. Originally it was found that loss of the P-light chain did not affect the ATPase activity of myosin [4,5]. However, some recent reports have suggested that removal of the P-light chain does affect the interaction of myosin with actin. In skeletal muscle, removal of the P-light chain caused a change in the Ca^{2+} sensitivity of the myofibrillar ATPase [60], reflecting the ability of the P-light chain to stabilise the actin/myosin interaction in the presence of ATP. Removal of the P-light chain in cardiac myofibrils by selective proteolysis resulted in a three-fold increase in actomyosin ATPase activity [61]. The P-light chain contains a Ca^{2+}-binding site of high affinity [62,63], and it has been suggested [64] that this represents a second site of Ca^{2+} interaction in striated muscle, possibly concerned with control of cross-bridge movement. Although this view has been challenged [65], it is probable that the P-light chain will have an important role in the regulation of contraction.

The first report of the phosphorylation of the P-light chain was from the laboratory of Perry, who found that impure preparations of phosphorylase kinase catalysed the incorporation of phosphate into either a light chain fraction from skeletal muscle myosin, or intact myosin itself [66]. Significantly, the P-light chain contained endogenous phosphate when isolated from skeletal muscle. Sequence analysis showed that a serine residue was phosphorylated, but that the primary sequence around the serine was quite different to that found in phosphorylase. Subsequently, it was realised that the phosphorylation was not being catalysed by phosphorylase kinase, but by a contaminant enzyme which has since been purified and shown to be a specific light-chain kinase [8]. This was shown to be activated by physiological (1 μM) concentrations of Ca^{2+} [8].

The light chain kinase has now been highly purified from a number of muscle types and is well characterised. The basic structure is of a catalytic subunit which varies in molecular weight from 80 000 to 105 000 depending on the muscle [67–71], and a regulatory subunit, of molecular weight 17 000 [68,70–72]. This regulatory subunit is identical with calmodulin (formerly termed the calcium dependent regulator protein), which has been implicated in the regulation of a number of cytoplasmic and membrane bound enzymes. Calmodulin was initially discovered as a heat stable protein which activated a brain cyclic nucleotide phosphodiesterase in the presence of low concentrations of Ca^{2+} [73,74]. It was subsequently shown to be necessary for the Ca^{2+} stimulation of brain adenylate cyclase [75] and for maximal activation of the Ca^{2+}-dependent ATPase of erythrocytes [76]. Calmodulin is also a subunit of phosphorylase kinase [77], whose activity is dependent on low concentrations of Ca^{2+} [78]. This last enzyme is unusual in that calmodulin is firmly attached to the kinase even in the absence of Ca^{2+}, whereas for all the other systems involving calmodulin, it only associates with the catalytic protein in the presence of Ca^{2+} (see Chapter 1).

Calmodulin has a large degree of sequence homology with troponin-C [79], and appears to have similar Ca^{2+} affinities and binding sites. A general principle seems to be emerging that all enzymes or systems in the cytoplasm that are affected by μM concentrations of Ca^{2+} have calmodulin (or homologous proteins) as the component which binds Ca^{2+}. Of interest is the fact that mitochondria do not appear to contain calmodulin (B. Kemball and P.J. England, unpublished observations).

The light chain kinase purified from skeletal muscle is specific for myosin P-light chain [8], and it does not phosphorylate phosphorylase, phosphorylase kinase or histone [80]. It was reported that a Ca^{2+}-dependent protein kinase can be prepared from skeletal muscle which not only phosphorylated P-light chain, but also phosphorylase kinase and histone [81]. This claim has now been withdrawn (J.H. Wang, personal communication to the editor). Phosphorylase kinase not only contains calmodulin as a subunit but is also activated by the binding of a second molecule of calmodulin to the enzyme [82]. Since phosphorylase kinase phosphorylates itself, the failure to carry out control incubations in which myosin kinase is omitted is the explanation for the previous results. Muscle does *not*, therefore, contain a general Ca^{2+}-dependent protein kinase analogous to the cyclic AMP-dependent protein kinase (see Chapter 11). The activity of P-light chain kinase was highest in fast-twitch, white skeletal muscle, and lower in slow twitch skeletal and heart muscle [7]. In addition, muscle also contains a highly specific light chain phosphatase which removes the phosphate from the P-light chain [83].

In striated muscle, the purpose of P-light chain phosphorylation is unknown. With isolated myosin, phosphorylation had no effect on the steady-state rate of myosin ATPase in the absence or presence of actin [83]. However, these experiments do not rule out the possibility of a change in one of the enzymic steps during the ATPase reaction, or that phosphorylation causes a change in cross-bridge orientation which will not be detected using solubilised systems. Phosphorylation did not change the affinity of whole myosin for Ca^{2+} [84]. However, it has been reported that phosphorylation decreased the affinity of isolated P-light chain for Ca^{2+} [85]. In any case, P-light chain phosphorylation has been shown to occur in vivo. In both frog and rabbit skeletal muscles in situ, the phosphate content of the P-light chain was shown to increase from 0.5 mol phosphate/mol light chain at rest to 0.9 mol/mol after a short tetanic contraction. This indicates that when there is a rise in intracellular Ca^{2+} phosphorylation of the P-light chain can occur.

In heart muscle, the situation is rather more confused. It was originally reported [87] that in unstimulated perfused rabbit heart the P-light chain was fully phosphorylated, and that this decreased when the heart was perfused with adrenaline. This result was unexpected, as adrenaline is known to increase intracellular Ca^{2+} in the heart (see Section 3.1), and this would be expected to stimulate light chain kinase. In addition, the maximal phosphorylation of P-light chain in unstimulated heart was not consistent with the relatively low level of light-chain kinase found in heart [30]. However, with perfused rat heart this result was not repeated [88]. It was found that with unstimulated hearts there was only a low level of P-light chain phosphorylation, and this was not changed with adrenaline. Recently the original observation [87] has been questioned by the authors [89], and it appears that P-light chain phosphorylation is less important in heart than in skeletal muscle.

From the above discussion it is clear that in striated muscle P-light chain phosphorylation is unlikely to be of direct regulatory significance in the control of muscle contraction. In smooth muscle and non-muscle contractile systems, in contrast, current evidence points to P-light chain phosphorylation being an obligatory step in the activation of contraction [9]. It was first observed with platelet myosin [90] that phosphorylation of the P-light chain caused an increase in the actin-activated myosin ATPase. More recent work has shown that this activation is not restricted to platelets, but occurs with a number of non-muscle contractile systems [10], and also in smooth muscle. Generally, myosin or actomyosin isolated from various smooth muscles showed a very low ATPase activity, which was not stimulated by Ca^{2+} [91–93]. Incubation of these preparations with myosin light chain kinase and ATP-Mg resulted in phosphorylation of the P-light chain, and a large increase in the actomyosin ATPase activity [93]. The actomyosin preparations with low ATPase activity were found to have dephosphorylated P-light chains [91–93]. By using an actomyosin preparation with added light chain kinase, it was shown that the ATPase activity was directly proportional to the amount of light chain in the phosphorylated form, and that this was controlled by the Ca^{2+} concentration in the incubation [93]. These results have led to the following scheme being proposed for control of smooth muscle contraction [9]. When the muscle is relaxed, the cytoplasmic concentration of Ca^{2+} is low, the light chain kinase is inactive, and the P-light chain is in the dephosphorylated form. There is, therefore, no actin/myosin interaction (i.e., relaxed muscle). When the muscle is stimulated, the cytoplasmic Ca^{2+} increases to 10^{-6}–10^{-5}M, leading to an activation of the kinase and phosphorylation of the P-light chain. The myosin can now interact with the actin filaments, resulting in contraction. Relaxation will occur when the Ca^{2+} concentration falls below that necessary to activate the kinase, allowing the phosphatase to dephosphorylate the P-light chain. At present it is not known if the activity of the phosphatase is controlled during the contraction/relaxation cycle, or if it remains active throughout.

Recent evidence in vivo also supports this scheme [94,95]. Pig artery strips were frozen in either a relaxed or contracted state [94], and the proportion of P-light chain in the phosphorylated form estimated by 2-dimensional gel electrophoresis. In the relaxed muscle there appeared to be very little P-light chain phosphorylation, but in the contracted muscle approximately 30% of the light chain was in the phosphorylated form. Use of skinned ileum or gizzard strips [95] also showed a correlation between P-light chain phosphorylation and tension developed when the Ca^{2+} concentration was varied.

In general, contraction in smooth muscle is stimulated by increases in intracellular free Ca^{2+}, while relaxation is often associated with increases in cyclic AMP. It has recently been reported [96] that the catalytic subunit of the light chain kinase from turkey gizzard can be phosphorylated by cyclic AMP-dependent protein kinase, and that this resulted in a 2-fold decrease in light chain kinase activity. If this were to occur in vivo, it would be one mechanism whereby increases in cyclic AMP could directly cause relaxation.

There is evidence that Ca^{2+} can affect contraction in smooth muscle by mechanisms in addition to that involving light-chain phosphorylation. Actomyosin from vas deferens was prepared with the P-light chain fully phosphorylated, but free of light chain kinase and phosphatase [92]. Although Ca^{2+} did not cause a change in P-light chain phosphorylation with this preparation, there was an increase in actomyosin ATPase. This suggested the possibility of Ca^{2+} acting in addition via a troponin-like system in smooth muscle. Ebashi has reported [24] the presence in turkey gizzard of two proteins which restored Ca^{2+} sensitivity to smooth muscle actomyosin, and these appeared to have certain properties similar to troponin. Moreover, Ebashi has put forward a contrary view to that above concerning the role of P-light chain phosphorylation in the control of smooth muscle contraction [25,26]. He had claimed that it is possible to dissociate activation of actomyosin ATPase from P-light chain phosphorylation. Incubation of gizzard actomyosin with Ca^{2+} resulted in phosphorylation of the P-light chain, and actin/myosin interaction as measured by superprecipitation. Removal of Ca^{2+} caused a reversal of actin/myosin interaction, but the P-light chain remained phosphorylated. If the same experiment was performed in the presence of excess light chain phosphatase, there was still Ca^{2+}-stimulated superprecipitation, but no P-light chain phosphorylation. These results suggest that although phosphorylation of P-light chain occurs readily in the presence of Ca^{2+}, it is not obligatory for contraction, and that control of contraction occurs via troponin-like proteins on the thin filament.

At the present time it is difficult to reconcile these different experiments and interpretations. It may be that superprecipitation does not measure the same phenomenon as ATPase activity, and that the former is independent of P-light chain phosphorylation. It is clear that a considerable amount of work is required before a full picture of the control of smooth muscle contraction emerges.

One problem only partially resolved concerns the control of myosin light-

'chain kinase in non-muscle cells. The kinase from platelets, myoblasts, macrophages and Hela cells was originally reported [97] to be insensitive to Ca^{2+}. However, the recent discovery that calmodulin is a part of muscle light chain kinase [68,70–72] has led to the reinvestigation of the kinase from non-muscle cells [10], and it appears that calmodulin confers Ca^{2+} sensitivity on this also. It may therefore be that the same mechanism for control of contractile activity operates in non-muscle cells and in smooth muscle.

2.3. Other contractile proteins

Apart from troponin and myosin light chains, a number of other contractile 'proteins have been reported to be phosphorylated, although in general the role of phosphorylation is unclear.

When live frogs were injected with $^{32}P_i$ and left for several days, tropomyosin isolated from leg muscles was found to contain ^{32}P as phosphoserine [98]. The amount of phosphate incorporated was approximately 0.3 mol/mol protein. It had also been reported from the same laboratory [99] that tropomyosin could be phosphorylated in chicken muscle. The phosphate is attached to a specific site (serine 283) in the protein [89]. The kinase responsible for this phosphorylation is at present unknown. Tropomyosin is not phosphorylated by phosphorylase kinase [14] or cyclic AMP-dependent protein kinase [56] under conditions where troponin-I is phosphorylated.

Filamin is a protein of molecular weight 240 000 which is found in large quantities in smooth muscle [100–102], and also in other non-muscle cells. It is a protein which binds actin, and can inhibit actin activation of myosin ATPase [101,103,104]. It was originally observed [105] in fibroblasts incubated with $^{32}P_i$ that a protein of molecular weight 240 000, which was precipitated with antibody to filamin, was phosphorylated. It was also shown that filamin from chicken gizzard could be phosphorylated by incubation with cyclic AMP-dependent protein kinase. Phosphorylation of filamin in intact mammalian smooth muscle has also been reported [106], and purified filamin from this source can also be phosphorylated by cyclic AMP-dependent protein kinase. Using cell fractions, filamin phosphorylation was stimulated by cyclic AMP, but not by cyclic GMP which stimulated a different group of proteins [106]. Both in vivo and in vitro the amount of phosphate in filamin 'was 0.2–0.5 mol/mol filamin. As yet, however, there is no evidence that phosphorylation of filamin affects its interaction with actin or other contractile proteins.

There is also evidence that the heavy chain of myosin contains covalently bound phosphate. Cardiac myosin isolated from animals injected with $^{32}P_i$

[107], or from cultures of foetal cardiac cells [108], was found to contain covalently bound ^{32}P. The turnover of ^{32}P was the same as that of the myosin heavy chain [108], indicating that the phosphate was probably incorporated during or immediately following protein synthesis, and then remained permanently attached for the lifetime of the myosin. This phosphate was shown to be linked covalently to serine and threonine residue, and appeared to be located in only one region of the molecule [109]. The exact location of this phosphate or its significance is not known at present.

3. Phosphorylation of membrane proteins

3.1. Cardiac muscle

The phosphorylation of membranes from cardiac muscle has been intensively studied for a number of years, and a large amount of information is available from in vitro experiments. Contraction in the heart is affected by several hormones and a large number of pharmacological agents, many of which could cause their effects via changes in membrane protein phosphorylation. For these reasons cardiac muscle will be discussed first in this section, and phosphorylation of membrane proteins in skeletal muscle will be discussed after.

Before discussing phosphorylation of membrane proteins, a brief summary will be given of the changes in Ca^{2+} concentration and fluxes during a cycle of contraction and relaxation [34,58,110,111]. When the cell membrane is depolarised at the end of diastole, Ca^{2+} channels open in the cell membrane, and Ca^{2+} enters the cell during the plateau phase of the action potential. A larger amount of Ca^{2+} is also released from the sarcoplasmic reticulum. There is some debate whether this release is triggered directly by the cell membrane depolarisation, or by the Ca^{2+} entering from the extracellular medium, but the overall result is to increase the free cytoplasmic Ca^{2+} from approximately 0.1 μM during diastole to 10 μM during systole. However, as the intracellular concentration of troponin is approximately 50 μM, and as each troponin can bind two Ca^{2+} ions, to raise the free Ca^{2+} concentration to 10 μM requires a total change in cytoplasmic Ca^{2+} of at least 0.1 $\mu mol/ml$ intracellular volume [112]. Relaxation occurs by the uptake of Ca^{2+} back into the sarcoplasmic reticulum by the action of the Ca^{2+} dependent ATPase. In addition, the Ca^{2+} that originally crossed the cell membrane during systole has to be transported back out again. It is generally thought that this occurs by Na^+/Ca^{2+} exchange

[58], although there is evidence of a Ca^{2+}-dependent ATPase mechanism [113,114].

Adrenaline and other similar agents cause both an increase in the force of contraction in the heart, and a decrease in the relaxation time [58]. Present evidence suggests that the increase in force is brought about by an increased influx of Ca^{2+} across the sarcolemma during each beat, which progressively leads to an increase in the total cell Ca^{2+} [58, 115–117]. The decreased relaxation time appears to be related to an increased rate of uptake of Ca^{2+} into the sarcoplasmic reticulum [34,58] (and possibly increased transport to the intracellular medium). In general, these changes in contractility are preceded by a rise in the intracellular concentration of cyclic AMP [118,119], suggesting that protein phosphorylation could be important in causing changes in Ca^{2+} fluxes.

It was first reported in 1969 that Ca^{2+} uptake into a preparation of sarcoplasmic reticulum vesicles from heart muscle was stimulated by cyclic AMP [120]. This was observed in the absence or presence of oxalate, which is used experimentally to precipitate Ca^{2+} inside the vesicles and enhance Ca^{2+} uptake. Subsequent work, much of it by the group of Katz, has led to a detailed understanding of this stimulation by cyclic AMP. It was shown that a membrane-bound protein in sarcoplasmic reticulum vesicles, of molecular weight 22 000, could be phosphorylated by endogenous and exogenous cyclic AMP-dependent protein kinase [121–123]. This protein was named 'phospholamban' by Katz. Phosphorylation was shown to increase oxalate-dependent Ca^{2+} uptake, and Ca^{2+}-dependent ATPase activity of isolated vesicles [124,125]. Dephosphorylation of phospholamban by treatment of the vesicles with phosphoprotein phosphatase resulted in a decrease in the rate of Ca^{2+} uptake [126]. It has been suggested [127] that phosphorylation of phospholamban in vivo, stimulated by adrenaline, would lead to a more rapid uptake of Ca^{2+} during diastole, resulting in the decrease in relaxation time observed with catecholamines. There is also some evidence that a Ca^{2+}-stimulated protein kinase could be involved in the phosphorylation of phospholamban [128], although this has yet to be substantiated.

Phospholamban is an entity quite distinct from the Ca^{2+}-dependent ATPase, which has a molecular weight of 90 000. It is a very hydrophobic protein, intrinsic to the sarcoplasmic reticulum membrane, although the part of the protein containing the phosphorylated site must be exposed to the aqueous phase. It is proposed that phospholamban interacts with the ATPase, and stimulates the ATPase when phosphorylated [58].

The studies reported above were all carried out measuring the oxalate-dependent Ca^{2+} uptake over a time scale of several minutes. However, Ca^{2+}

uptake in the beating heart occurs in 50–200 ms, and the oxalate-dependent uptake can only account for a small part of this [112]. Using rapid kinetic techniques, it has been shown that cardiac sarcoplasmic-reticulum vesicles can rapidly accumulate Ca^{2+} in the absence of oxalate. This rate is sufficient to account for the uptake of Ca^{2+} during diastole [129–131]. As this rapid uptake is probably more closely related to the physiological process than that observed with oxalate, the effects of phosphorylation of membrane protein on this rapid uptake are of obvious importance. The initial rate of Ca^{2+} uptake '(measured after 300 msec) into sarcoplasmic reticulum vesicles was increased by 50% after phosphorylation with cyclic AMP-dependent protein kinase [34]. Although this increase is not as great as that seen with oxalate, it would still be sufficient to account for most of the decrease in relaxation time caused by adrenaline.

All of the above experiments have been carried out in vitro with only partially pure preparations of sarcoplasmic reticulum. To date it has not been demonstrated that phosphorylation of phospholamban (or any other sarcoplasmic reticulum protein) occurs in vivo. One of the important criteria, originally suggested by Krebs [132] and since modified by Krebs and others [45,133], to decide if phosphorylation of a protein is part of a hormonal response, is to observe phosphorylation of the protein in vivo in response to the hormone. This phosphorylation must precede, or occur at the same time, as the physiological response to the hormone. Problems with the hydrophobic nature of phospholamban, its relatively low concentration in heart, and difficulty of isolation have meant that in vivo phosphorylation has yet to be demonstrated. Until this has been done the hypothesis that adrenaline causes a decrease in the relaxation time of the heart by phosphorylation of phospholamban remains speculative.

Vesicles which appear to be predominantly sarcolemmal in origin can be prepared from heart by similar methods to those used for preparation of sarcoplasmic reticulum vesicles. These will also accumulate Ca^{2+} in the presence of oxalate in an ATP-dependent process [114]. When these vesicles were incubated with ATP and cyclic AMP, there was an increase in Ca^{2+} uptake. This could also be stimulated by exogenously added protein kinase. A number of proteins appeared to be phosphorylated under these circumstances, with molecular weights varying from 11 000 to 95 000 [135,136]. Evidence has been presented that sarcolemma vesicles as prepared are inside-out, i.e., Ca^{2+}-uptake into the vesicles is equivalent to removal of Ca^{2+} from the cytoplasm in the intact cell [137]. Any increase in uptake on phosphorylation 'would therefore be equivalent to an increased rate of removal of Ca^{2+} from the cytoplasm. Ca^{2+} uptake in the absence of oxalate was also stimulated by treatment with cyclic AMP and protein kinase [138]. Although the mechanism whereby phosphorylation increases Ca^{2+} uptake is unknown [34], this could represent another mechanism for the more rapid removal of Ca^{2+} from the cytoplasm when the heart is stimulated by adrenaline.

In addition to the above mechanisms for increasing the rate of Ca^{2+} efflux during diastole, there is evidence that Ca^{2+} influx during systole can be increased with adrenaline. It has been postulated that an increase in cyclic AMP causes an increase in Ca^{2+} influx into the heart through the so-called 'slow Ca^{2+} channels' [139–141]. This would then lead to an increased cytoplasmic Ca^{2+} concentration which would result in an increase in the force of contraction. Although there is no direct evidence that this is brought about by protein phosphorylation, it has been suggested [34] that the protein of molecular weight 11 700 which is phosphorylated in sarcolemma preparations could be the slow Ca^{2+} channel.

There is one report of phosphorylation of sarcolemmal proteins in vivo [142]. Rat hearts were perfused with $^{32}P_i$, and following adrenaline administration were freeze clamped. Sarcolemmal vesicles were prepared from the frozen hearts, and the proteins separated by polyacrylamide gel electrophoresis. Two proteins of molecular weights 36 000 and 27 000 were found to be phosphorylated, but the phosphorylation of the 27 000 molecular weight protein only was increased by adrenaline. This protein could possibly be the same as a protein of 24 000 which was reported to be phosphorylated in isolated sarcolemma preparations by cyclic AMP-dependent protein kinase [143]. The protein was phosphorylated in hearts perfused with adrenaline with a slower time course than the increase in contraction. The increase in contraction had reached a maximum before a significant increase in phosphorylation of the protein had occurred. This crucial observation suggests that phosphorylation of this protein cannot be the mechanism whereby adrenaline causes an increase in cytoplasmic Ca^{2+}. It is obvious that more studies of this type are required before the hypothesis that phosphorylation of sarcolemmal proteins can mediate the changes in Ca^{2+} fluxes seen in the heart with adrenaline can be accepted.

3.2. Skeletal muscle

Catecholamines have a much less pronounced effect on the force of contraction in skeletal muscle as compared to their effect in cardiac muscle [144]. However, there is a definite increase in the force of contraction when skeletal muscle is treated with adrenaline before stimulation as well as a decrease in the relaxation time in slow twitch muscle. As with cardiac muscle, phosphorylation of membrane proteins is implicated in this effect.

It was initially observed [145] that incubation of sarcolemma from rabbit skeletal muscle with cyclic AMP-dependent protein kinase resulted in an increase in the rate of ATP-dependent Ca^{2+} uptake in the presence of oxalate. This was accompanied by incorporation of phosphate into protein. Similar preparations from rat were found to contain an endogenous protein kinase which phosphorylated three proteins with molecular weights below 30 000. As most of the Ca^{2+} for contraction in skeletal muscle is thought to arise from the sarcoplasmic reticulum, however, the significance of changes in sarcolemmal Ca^{2+} fluxes is not clear.

Phosphorylation of sarcoplasmic reticulum vesicles, associated with changes in Ca^{2+} fluxes, has been reported for preparations from slow twitch ·muscles [145,146]. Incubation of sarcoplasmic reticulum with cyclic AMP-dependent protein kinase [145,146] or phosphorylase kinase [146] resulted in an increase of 20–50% in the rate of Ca^{2+} uptake in the presence of oxalate. This was much less than the increase observed with cardiac sarcoplasmic reticulum. Phosphorylation of a 22 000 molecular weight protein, presumably phospholamban or a similar protein, occurred under these conditions. This may be associated, as in heart muscle, with the decrease in relaxation time observed with adrenaline. With sarcoplasmic reticulum from fast skeletal muscle, in contrast, there was no phosphorylation of a 22 000 molecular weight protein [145,146]. However, one group observed [146] an increase in oxalate-dependent Ca^{2+} uptake on incubation with cyclic AMP-dependent protein kinase, whereas no increase in uptake was reported [146] by another group. It has also been reported [147] that a Ca^{2+}-activated protein kinase (possibly phosphorylase kinase) is present in preparations of skeletal muscle sarcoplasmic reticulum. Surprisingly, when this kinase was activated by Ca^{2+} and ATP, there was a decrease in the rate of Ca^{2+} uptake by the vesicles. There appear to be a number of contradictory findings with regard to the effects of phosphorylation on sarcoplasmic reticulum from fast skeletal muscle which cannot be reconciled at the present time.

4. Concluding remarks

There are now an appreciable number of contractile and membrane proteins in muscle that are known to be phosphorylated by protein kinases. Some of these, in particular the myosin P-light chain, appear to be phosphorylated as part of the normal contraction process. This occurs because the kinase catalysing the phosphorylation is Ca^{2+} dependent, and will therefore be activated when Ca^{2+} is released in response to membrane depolarisation.

Many of the proteins are phosphorylated by cyclic AMP-dependent protein kinase, however, and phosphorylation of these proteins is therefore only likely to occur in response to hormonal stimulation of adenylate cyclase. These latter phosphorylations, e.g. troponin-I and phospholamban, probably have a modulatory role, rather than being an obligatory part of contraction. The Ca^{2+}-stimulated phosphorylations probably occur in all muscle types, although their regulatory significance is not always clear. In contrast, the cyclic AMP-dependent phosphorylations occur only in selected muscle types. This is a direct reflection of the fact that contraction in certain muscle types, particularly cardiac and smooth muscle, is readily affected by hormones. However, these two types of phosphorylation should not be thought of as being independent of each other, since as was discussed above, cyclic AMP-dependent phosphorylations can bring about increases in cytoplasmic Ca^{2+}.

There is now good evidence for the phosphorylation of many muscle proteins in vitro and/or in vivo. With troponin-I and myosin P-light chain, phosphorylation has been shown in vivo and in vitro, and modifications of function have been demonstrated in vitro. Although there are still certain aspects of the phosphorylation of these proteins which are not completely understood, they represent two of the better documented examples of this phenomenon. There is far less information available on the other contractile proteins which have been reported to be phosphorylated. In some cases the 'phosphorylations may have a regulatory function, whereas in other cases a structural role may be indicated. Phosphorylation of membrane proteins, associated with changes in Ca^{2+} fluxes, is well documented in vitro, but almost no experiments in vivo have been reported. This is an important area where a number of highly significant observations still have to be made.

During the last ten years it has become clear that muscle contraction, in common with many other cell processes, is regulated by protein phosphorylation/dephosphorylation. At the moment there are many areas which are only partially understood. The next ten years will doubtless see a clarification of this whole field, so that we will have as much understanding of the role of protein phosphorylation in muscle contraction as we now have of its role in the control of glycogen metabolism.

References

1. Huxley, A.F. and Neidergerke, R. (1954) Nature 173, 971–973.
2. Lowey, S., Slayter, M.S., Weeds, A.G. and Baker, M. (1969) J. Mol. Biol., 42, 1–29.
3. Weeds, A.G. and Frank, G. (1972) Cold Spring Harbor Symp., 37, 9–17.
4. Gazith, J., Himmelfarb, S. and Harrington, W.F. (1970) J. Biol. Chem., 245, 15–22.

5. Weeds, A.G. and Lowey, S. (1971) J. Mol. Biol., 61, 701–725.
6. Frearson, N. and Perry, S.V. (1975) Biochem. J., 151, 99–107.
7. Frearson, N., Focant, B.W.W. and Perry, S.V. (1976) FEBS Lett., 63, 27–32.
8. Pires, E., Perry, S.V. and Thomas, M.A.W. (1974) FEBS Lett., 41, 292–296.
9. Adelstein, R.S. (1978) Trends in Biochem. Sci. 3, 27–30.
10. Adelstein, R.S., Scordilis, S.P. and Trotter, J.A. (1979) Meth. Achiev. Exp. Pathol., 8, 1–41.
11. Mihashi, K. (1964) Arch. Biochem. Biophys., 107, 441–448.
12. Elzinga, M., Collins, J.H., Kuehl, W.M. and Adelstein, R.S. (1973) Proc. Natl. Acad. Sci. USA, 70, 2687–2691.
13. Hanson, J. and Lowey, S. (1963) J. Mol. Biol., 6, 46–60.
14. Elzinga, M. and Lu. R.C. (1976) in: Contractile Systems in Non-Muscle Tissues (Perry, S.V., Margreth, A. and Adelstein. R.S., eds.), pp. 29–37, North-Holland, Amsterdam.
15. Elzinga, M., Maron, B.J. and Adelstein, R.S. (1976) Science, 191, 94–95.
16. Lu. R., and Elzinga, M. (1976) Cold Spring Harbor Conf., Cell Proliferation, 3, 487–490.
17. Lymm, R.W. and Taylor, E.W. (1971) Biochemistry, 10, 4617–4624.
18. Huxley, A.F. (1971) Proc. Roy. Soc. B., 178, 1–27.
19. Ebashi, S., Endon, M. and Ohtsuki, I. (1969) Quart. Rev. Biophys., 2, 351–384.
20. Huxley, H.E. (1972) Cold Spring Harbor Symp., 37, 361–376.
21. Ebashi, S. and Endo, M. (1968) Prog. Biophys. Mol. Biol., 18, 125–183.
22. Perry, S.V. Cole, H.A., Frearson, N. Moir, A.J.G. Morgan, M. and Pires, E. (1976) in: Molecular Basis of Motility (Heilmeyer L.M.G., Ruegg J.C., Wieland T., eds.), pp. 107–121, Springer-Verlag, Berlin.
23. Potter, J.D., Nagy, B., Collins, J.H., Seidel, J.C., Leavis, P., Lehrer, S.S. and Gergely, J. (1976) in: Molecular Basis of Motility (Heilmeyer L.M.G., Ruegg J.C. and Wieland, T., eds.), pp. 93–106, Springer-Verlag, Berlin.
24. Ebashi, S., Toyo-oka, T. and Nonomura, Y. (1975) J. Biochem., 78, 859–861.
25. Mikawa, T., Nonomura, Y. and Ebashi, S. (1977) J. Biochem., 82, 1789–1791.
26. Mikawa, T., Toyo-oka, T., Nonomura, Y. and Ebashi, S. (1977) J. Biochem. 81, 273–275.
27. Kendrick-Jones, J., Lehman, W. and Szent-Gyorgyi, A.G. (1970) J. Mol. Biol., 54, 313–326.
28. Kendrick-Jones, J., Szentkiralyi, E.M. and Szent-Gryorgyi, A.G. (1972) Cold Spring Harbor Symp., 37, 47–53.
29. Frearson, N. and Perry, S.V. (1975) Biochem. J., 151, 99–107.
30. Frearson, N. Focant, B.W.W. and Perry, S.V. (1976) FEBS Lett., 63, 27–32.
31. Inesi, G. and Malan, N. (1976) Life Sci., 18, 773–780.
32. Hasselback, W. and Makinose, M. (1961) Biochem. Z., 333, 518–531.
33. Ebashi. S. and Lipmann, F. (1962) J. Cell. Biol., 14, 389–400.
34. Wollenberger, A. and Will, H. (1978) Life Sci., 22, 1159–1178.
35. Kuriyama, M. Osa, T. Ito, Y. and Suzuki, M. (1976) Adv. Biophys., 8, 115–190.
36. Greaser, M.L. and Gergely, J. (1971) J. Biol. Chem., 246, 4226–4233.
37. Bailey, C. and Villar-Palasi, C. (1971) Fed. Proc., 30, 1147.
38. Stull, J.T., Brostrom, C.O. and Krebs, E.G. (1972) J. Biol. Chem., 247, 5272–5274.
39. England, P.J., Stull, J.T., Huang, T.S. and Krebs, E.G. (1973) Metabolic Interconversions of Enzymes, 3, 175–184.
40. England, P.J., Stull, J.T. and Krebs, E.G. (1972) J. Biol. Chem., 247, 5275–5278.
41. Perry, S.V. and Cole, H.A. (1973) Biochem. J., 131, 425–428.
42. Perry, S.V. and Cole, H.A. (1974) Biochem. J., 141, 733–743.
43. Stull, J.T. and High, C.W. (1977) Biochem. Biophys. Res. Commun., 77, 1078–1083.
44. Bylund, D.B. and Krebs, E.G. (1975) J. Biol. Chem., 250, 6355–6361.
45. Nimmo, H.G. and Cohen, P. (1977) Adv. Cyclic Nucleotide Res., 8, 145–266.
46. Dobrovol'skii, R.B., Gusev, V.B., Martynov, A.V. and Severin E.S. (1976) Biokhimiya, 41, 1291–1296.

47. Gusev., N.B., Dobrovol'skii, A.B. and Severin, E.S. (1977) Biokhimiya, 43, 365–372.
48. Reddy, Y.S., Ballard, D., Giri, N.Y. and Schwartz, A. (1973) J. Mol. Cell. Cardiol., 5, 461–471.
49. Grand, R.J.A., Wilkinson, J.M. and Mole, L.E. (1976) Biochem. J., 159, 633–641.
50. Moir, A.J.G. and Perry, S.V. (1977) Biochem. J., 167, 333–343.
51. England, P.J. (1975) FEBS Lett., 50, 57–60.
52. England, P.J. (1976) Biochem. J., 160, 295–304.
53. Solaro, R.J., Moir, A.J.G. and Perry, S.V. (1976) Nature, 262, 615–617.
54. Ezrailson, E.G., Potter, J.D., Michael, L. and Schwartz, A. (1977) J. Mol. Cel. Cardiol., 9, 693–698.
55. England, P.J. (1977) Biochem. J., 168, 307–310.
56. Ray, K.P. and England, P.J. (1976) FEBS Lett., 70, 11–16.
57. Reddy, Y.S. and Wyborny, L.E. (1976) Biochem. Biophys. Res. Commun., 73, 703–709.
58. Katz, A.M. (1977) The Physiology of the Heart, Raven Press, New York.
59. Buss, J.E. and Stull, J.T. (1977) FEBS Lett., 73, 101–104.
60. Pemrick, S.M. (1977) Biochemistry, 16, 4047–4054.
61. Malhotra, A., Huang, S. and Bhan, A. (1979) Biochemistry, 18, 461–467.
62. Werber, M.M. Gaffin, S.L. and Oplatka, A. (1972) J. Mechanochem. Cell. Motility, 1, 91–96.
63. Morimoto, K. and Harrington, W.F. (1974) J. Mol. Biol., 88, 693–709.
64. Haselgrove, J.C. (1975) J. Mol. Biol., 92, 113–143.
65. Bashaw, C.R. and Reed, G.M. (1977) FEBS Lett., 81, 386–390.
66. Perrie, W.T., Smillie, L.B. and Perry, S.V. (1973) Biochem. J., 135, 151–164.
67. Daniel, J. and Adelstein, R.S. (1976) Biochemistry, 15, 2370–2377.
68. Yazawa, M. and Yagi, K. (1977) J. Biochem., 82, 287–289.
69. Pires, E.M.V. and Perry, S.V. (1977) Biochem. J., 167, 137–146.
70. Dabrowska, R., Aromatorio, D., Sherry, J.M.F. and Hartshorne, D.J. (1977) Biochem. Biophys. Res. Commun., 78, 1263–1272.
71. Nairn, A.C. and Perry, S.V. (1978) Biochem. J., 179, 89–97.
72. Dabrowska, R., Sherry, J.M.F., Aromatorio, D.K. and Hartshorne, D.J. (1978) Biochemistry, 17, 253–258.
73. Cheung, W.Y. (1970) Biochem. Biophys. Res. Commun., 38, 533–538.
74. Kakiuchi, S., Yamazaki, R. and Nakajima, H. (1970) Proc. Japan. Acad., 46, 589–592.
75. Brostrom, C.Q., Huang, Y.C., Breckenridge, B.M. and Wolff, D.J. (1975) Proc. Natl. Acad. Sci. USA, 72, 64–68.
76. Gopinath, R.M. and Vicenzi, F.F. (1977) Biochem. Biophys. Res. Commun., 77, 1203–1209.
77. Cohen, P., Burchell, A., Foulkes, J.G., Cohen, P.T.W., Vanaman, T.C. and Nairn, A.C. (1978) FEBS Lett., 92, 287–293.
78. Brostrom, C.O., Hunkeler, F.L. and Krebs, E.G. (1971) J. Biol. Chem., 246, 1961–1967.
79. Amphlett, G.W., Vanaman, T.C. and Perry, S.V. (1976) FEBS Lett., 72, 163–167.
80. Nairn, A.C. and Perry, S.V. (1979) Biochem. J., 179, 89–97.
81. Waisman, D.M., Singh, T.J. and Wang, J.H. (1978) J. Biol. Chem., 253, 3387–3390.
82. Cohen, P., Picton, C. and Klee, C.B. (1979) FEBS Lett., 104, 25–30.
83. Morgan, M., Perry, S.V. and Ottoway, J. (1976) Biochem. J., 157, 687–697.
84. Holroyde, M.J., Potter, J.D. and Solaro, R.J. (1976) Biophys. J., 25, 243a.
85. Alexis, M.N. and Gratzer, W.B. (1978) Biochemistry, 17, 2319–2325.
86. Barany, K. and Barany, M. (1971) J. Biol. Chem., 252, 4752–4754.
87. Frearson, N., Solaro, R.J. and Perry, S.V. (1976) Nature, 264, 801–802.
88. England, P.J., Ray, K.P., Hibberd, M.G., Jeacocke, S.A., Murray, K.J. and Hollinworth, D.N. (1979) Horm. Cell. Regul., 3, 99–114.
89. Perry, S.V., Cole, H.A., Frearson, N., Moir, A.J.G. Nairn, A.C. and Solaro, R.J. (1979) FEBS Meeting – 1978, in press.

172

90. Adelstein, R.S. and Conti, M.A. (1975) Nature, 256, 597–598.
91. Gorecka, A., Aksoy, M.A. and Hartshorne, D.J. (1976) Biochem. Biophys. Res. Commun., 71, 325–331.
92. Chacko, S., Conti, M.A. and Adelstein, R.S. (1977) Proc. Natl. Acad. Sci. USA, 74, 129–133.
93. Small, J.V. and Sobieszek, A. (1977) Eur. J. Biochem., 76, 521–530.
94. Driska, S.P. and Murphy. R.A. (1979) Biophys. J., 25, 73a.
95. Cassidy, P.S., Hoar, P.E. and Kerrick, W.G.L. (1979) Biophys. J., 25, 73a.
96. Adelstein, R.S., Conti, M.A., Hathaway, D.R. and Klee, C.B. (1978) J. Biol. Chem., 253, 8347–8350.
97. Adelstein, R.S., Chacko, S., Scordilis, S.P., Barylko, B., Conti, M.A. and Trotter, J.A. (1977) in: Calcium-binding proteins and Calcium Function (Wasserman, R.M., Corradino, R.A., Carafoli, E., Kretsinger, R.M., Machennan, D.M. and Seigel, F.L., eds.), pp. 251–261 North-Holland, Amsterdam.
98. Ribolow, H. and Barany, M. (1977) Arch. Biochem. Biophys., 179, 718–720.
99. Barany, M., Barany, K., Gaetjens, E. and Horvath, B.Z. (1974) in: Exploratory Concepts in Muscular Dystrophy (Milkorat, A.T., ed.), Vol. II, pp. 451–462, Excerpta Medica, Amsterdam.
100. Wang, K. Ash., J.F. and Singer, S.J. (1975) Proc. Natl. Acad. Sci. USA, 72, 4483–4486.
101. Shizuta, Y., Shizuta, H., Gallo, M., Davies, P., Pastan, I. and Lewis, M.S. (1976) J. Biol. Chem. 251, 6562–6567.
102. Wang, K. (1977) Biochemistry, 16, 1857–1865.
103. Wang, K. and Singer, S.J. (1977) Proc. Natl. Acad. Sci. USA, 74, 2021–2025.
104. Davies, P., Bechtel, P. and Pastan, I. (1977) FEBS Lett., 77, 228–232.
105. Davies, P., Shizuta, Y., Olden, K., Gallo, M. and Pastan, I. (1977) Biochem. Biophys. Res. Commun., 74, 300–307.
106. Wallach, D., Davies, P.J.A. and Pastan, I. (1978) J. Biol. Chem., 253, 4739–4745.
107. McPherson, J., Fenner, C., Smith, A., Mason, D.T. and Wickmann-Coffelt, J. (1974) FEBS Lett., 47, 149–154.
108. Andreasen, T., Castles, J.J., Saito, W.Y., Chacko, K., Fenner, C., Mason, D.T. and Wickman-Coffelt, J. (1975) Develop. Biol., 47, 366–375.
109. Fenner, C., Mason, D.T. and Wickman-Coffelt, J. (1977) Anal. Biochem., 78, 188–196.
110. Langer, G.A. (1973) Ann. Rev. Physiol., 35, 55–86.
111. Reuter, H. (1974) Circul. Res., 34, 599–605.
112. Solaro, R.J., Wise, R.M. Shiner, J.S. and Briggs, F.N. (1974) Circul. Res. 34, 525–530.
113. Stam, A., Wiglicki, W.B., Gertz, E.W. and Sonnenblick, E.M. (1973) Biochim. Biophys. Acta, 298, 927–931.
114. Sulakhe, P.V., Leung, N.L.-K. and St. Louis, P.J. (1976) Canad. J. Biochem., 54, 438–445.
115. Van Winkle, W.B. and Schwartz, A. (1976) Ann. Rev. Physiol., 38, 247–272.
116. Allen, D.G., Jewell, B.R. and Wood, E.M. (1976) J. Physiol., 254, 1–17.
117. Jewell, B.R. (1977) Circulation Res., 40, 221–230.
118. Sobel, B.E. and Mayer, S.E. (1973) Circulation Res., 32, 407–414.
119. Tsien, R.W. (1977) Adv. Cyclic Nucleotide Res., 8, 363–420.
120. Entman, M.L., Levey, G.S. and Epstein, S.E. (1969) Circulation Res., 25, 429–438.
121. Wray, M.L., Gray, R.R. and Olsson, R.A. (1973) J. Biol. Chem., 248, 1496–1498.
122. Kirchberger, M.A., Tada, M. and Katz, A.M. (1974) J. Biol. Chem., 249, 6166–6173.
123. Laraia, P.J. and Morkin, E. (1974) Circulation Res., 35, 298–306.
124. Kirchberger, M.A., Tada, M. Repke, D.I. and Katz, A.M. (1972) J. Mol. Cell. Cardiol., 4, 673–680.
125. Tada, M., Kirchberger, M.A., Repke, D.I. and Katz, A.M. (1974) J. Biol. Chem., 249, 6174–6180.
126. Tada, M., Kirchberger, M.A. and Li, H-C. (1975) J. Cyclic Nucleotide Res., 1, 329–338.
127. Katz, A.M., Tada, M. and Kirchberger, M.A. (1975) Adv. Cyclic Nucletoide Res., 5, 453–472.

128. Wray, H.L. and Gray, R.R. (1977) Biochim. Biophys. Acta, 461, 441–459.
129. Ebashi, S. and Endo, M. (1968) Progress Biophys. Mol. Biol., 18, 123–183.
130. Besch, M.R. and Schwartz, A. (1971) Biochem. Biophys. Res. Commun., 45, 286–292.
131. Will, H., Blanck, J., Smeltan, G. and Wollenberger, A. (1976) Biochim. Biophys. Acta, 449, 295–303.
132. Krebs, E.G. (1973) Endocrinology, Proc. 4th Internat. Cong., pp. 17–29, Excerpta Medica, Amsterdam.
133. Krebs, E.G. and Beavo, J.A. (1979) Ann. Rev. Biochem., 48, in press.
134. Hui, C.-W. Drummond, M. and Drummond, G.I. (1976) Arch. Biochem. Biophys., 173, 415–427.
135. Sulakhe, P.V. and St. Louis, P.J. (1976) Gen. Pharmacol., 7, 313–319.
136. Dowd, F.J., Pitts, B.J.R. and Schwartz, A. (1976) Arch. Biochem. Biophys., 175, 321–331.
137. Lullmann, H. and Peters, T. (1976) in: Recent Advances in Studies on Cardiac Structure and Metabolism (Roy, P.E. and Dhalla, N.S., eds.), Vol. 9, pp. 311–328, University Park Press, Baltimore.
138. Will, H., Schirpke, B. and Wollenberger, A. (1973) Acta. Biol. Med. Germany, 31, K45–52.
139. Schneider, J.A. and Sperelakis, N. (1975) J. Mol. Cell. Cardiol., 7, 249–273.
140. Watanabe, A.M. and Besch, H.R. (1974) Circulation Res., 35, 316–324.
141. Sperelakis, S. and Schneider, J.A. (1976) Am. J. Cardiol., 37, 1079–1085.
142. Walsh, D.A. Clippinger, M.S., Sivaramakrishnan, S. and McCullough, T.E. (1979) Biochemistry, 18, 871–877.
143. Krause, E.G., Will, M., Schirpke, B. and Wollenberger, A. (1975) Adv. Cyclic Nucleotide Res., 5, 473–490.
144. Robison, G.A., Butcher, R.W. and Sutherland, E.W. (1971) Cyclic AMP, Academic Press, New York.
145. Kirchberger, M.A. and Tada, M. (1976) J. Biol. Chem., 251, 725–729.
146. Schwartz, A., Entman, M.L., Kaniiki, K., Lane, L.K., Van Winkle, W.G. and Bornet, E.P. (1976) Biochim. Biophys. Acta., 426, 57–72.
147. Horl, W.H., Hennissen, M.P. and Heilmeyer, L.M.G. (1978) Biochemistry, 17, 759–766.

CHAPTER 8

Phosphorylation and the control of protein synthesis in reticulocytes

TIM HUNT

1. Introduction

Studies of the control of protein synthesis in rabbit reticulocytes have a long history. In 1956, Borsook and his collaborators at Caltech could say, justifiably, that 'The evidence argues for a mechanism in the reticulocyte that co-ordinates and equalises the rates of heme and globin synthesis' [1] and the following year they presented detailed studies showing that iron salts and transferrin stimulated the synthesis of globin in these cells [2]. Later results from London's and Rabinovitz's laboratories showed that the iron acted by allowing adequate synthesis of haemin, which was itself an effective stimulant of globin synthesis even in the presence of iron chelating agents [3–5]. They also found [6] that exogenous haemin inhibited endogenous haem synthesis, in accord with Borsook's argument. Detailed studies of the effect of iron or haemin on intact reticulocytes showed that the control of globin synthesis was fully reversible, and that in the absence of adequate supplies of haemin, inhibition of protein synthesis was accompanied by disaggregation of polysomes to 80S ribosomes, suggesting that haemin was necessary for the initiation of protein synthesis, since the polysome breakdown did not occur in the presence of elongation inhibitors of protein synthesis.

To discover the molecular basis of these effects required biochemical analysis of cell-free systems which were capable of making haemoglobin at a high rate, and which faithfully preserved the control mechanisms that operate in the intact reticulocyte. Such a system was described in 1964 by Lamfrom and Knopf [7], who seem to have been the first workers to appreciate the advantages of the crude lysate as a system for studying the synthesis of protein. This system was 10–20 times as active as the more highly fractionated cell-free systems then in use, and most significantly, it initiated the synthesis

Cohen (ed.) Recently discovered systems of enzyme regulation by reversible phosphorylation
© *Elsevier/North-Holland Biomedical Press, 1980*

176

of globin chains with relatively high efficiency, although the rate of initiation dropped to negligible levels after only a few minutes of incubation. The fact that the addition of haemin to this system could preserve the high rate of initiation seems to have been discovered inadvertantly by Zucker and Schulman [8] during studies of the pathway of assembly of haemoglobin using the lysate as the source of radioactively labelled nascent chains, although there had been a report the year before from Hammel and Bessman, who obtained stimulation of globin synthesis in a cell-free system from duck erythrocytes [9] by haemin. Publication of Zucker and Schulman's complete description of their work on protein synthesis [10] was confirmed and amplified by independent studies from Ed Herbert's laboratory [11]. Thus by 1968 it was possible to obtain cell-free systems from reticulocytes which synthesised protein at a high rate for periods up to an hour.

The question as to why initiation failed in the absence of haemin was partially answered by a simple mixing experiment done by Maxwell and Rabinovitz [12]. They found that ribosome-free supernatant (S100) from lysates incubated without haemin inhibited protein synthesis by fresh lysates, and that the appearance of this inhibitor, which they called the Haemin Controlled Repressor (HCR), was prevented by the presence of haemin. At the same time they found that ribosomes from inhibited incubations were as active as control ribosomes when recombined with fresh haemin-containing S100, a somewhat misleading result viewed in retrospect, though perfectly correct. At the time, it was widely suspected [13] that HCR might be globin or some derivative thereof, so that the control represented a simple feedback loop; one should remember that ideas of how initiation worked were extremely primitive in 1969, and it would have been quite impossible to suggest a plausible model of control when the pathway of initiation and the nature of the elements involved were unknown. Other ideas had been put forward. For example, Burka found that haemin inhibited ribonuclease [14], and thought that the inhibition of protein synthesis might be due to the action of the enzyme despite the clear evidence that the inhibition was reversible. Others suggested that haemin might play a part in the folding and release of nascent chains from the ribosomes [15], though it might have been thought that the absence of haemin should stabilise polysomes rather than the reverse if this were true. Several experiments with labelled haemin were made which suggested that haemin did not bind to nascent chains, although ribosomes were labelled [16]. Apart from the interest of these plausible old hypotheses, they serve to illustrate the danger of studies with haemin; it seems to bind to proteins readily, and has been shown to inhibit several enzymic reactions, almost always at unphysiologically high concentration.

Studies on HCR continued, and it was soon clear that it was a protein which was nothing to do with globin [17]. However, the next real advance in understanding in the haemin control system came from studies on the initiation of protein synthesis, and clarification of the series of steps leading to the formation of productive initiation complexes containing 80 S ribosomes, Met-tRNA$_f$ and mRNA. It gradually emerged that the first intermediates in the formation of these complexes were pre-initiation complexes containing native 40 S subunits, Met-tRNA$_f$ and GTP [18–20]. In the absence of haemin, these complexes rapidly disappear, before polysomes break down [21]. There is no decline in the levels of 40 S subunits, Met-tRNA$_f$ or GTP at any stage in the incubation; hence it follows that something prevents the formation of these complexes, and that something was most likely HCR. The initiation factor which catalysed the association of these components was identified as eIF-2, a trimeric protein with α, β and γ subunits with molecular weights of 36 000, 52 000 and 57 000 respectively [22]. It had been discovered that crude mixtures of initiation factors could overcome the inhibition caused by HCR, and in 1974 Clemens et al. reported that purified eIF-2 was the active component in this mixture [23]. Hence it seemed probable that HCR acted by inactivating eIF-2. Since purified HCR appeared to act catalytically [17], and since the inhibition was reversible, the conclusion that some form of reversible covalent modification of eIF-2 must be catalysed by HCR seems in retrospect inescapable, though it did not appear so at the time. To give an idea of possible alternatives canvassed, it could have been argued that HCR caused the accumulation of a low molecular weight inhibitor that interfered with eIF-2 function. We spent some time trying to detect such an entity, and studying the metabolism of ATP and GTP to see if any 'magic spot' like compounds could be detected. We never found any. Another idea was that HCR causes the accumulation of deacylated tRNA$_f^{met}$ on 40 S ribosomes by stimulating the activity of a met-tRNA$_f$ deacylase that could be detected on ribosomes [24]. Echoes of this hypothesis still reverberate [25], but I do not believe it correct for the following reasons. We obtained some ^{32}P-labelled tRNA$_f^{met}$ and added it to lysates together with [^{35}S]methionine in the presence and absence of haemin. We found that ^{32}P counts were lost from 40 S ribosomes at the same time as the methionine counts; there was no accumulation of deacylated initiator tRNA on the ribosomes. Another way of testing for the deacylation of met-tRNA$_f$ is to measure the rate of turnover of the methionine on this tRNA in the presence and absence of HCR. We know that overall levels of charged met-tRNA$_f$ are unaltered during inhibition by HCR; if rapid deacylation is occurring, the rate of turnover of the methionine in met-tRNA$_f$ should be relatively unaffected, or even increased after inhibition, whereas if deacylation does not occur, the rate of turnover of this methionine should be reduced in the same proportion as protein synthesis is inhibited. Such is the case; when protein synthesis is totally inhibited by an elongation inhibitor like sparsomycin, the rate of turnover of met-tRNA$_f$ falls to 1–5% of the control rate. When protein synthesis was inhibited by HCR to the extent of 75%, the rate of turnover of met-tRNA$_f$ was also reduced by 75%. Unfortunately, it is difficult to reconcile these results with those of Balkow et al. and Gross, whose experiments suggested the deacylation hypothesis, except to say that their experiments were less direct than ours.

The realisation that HCR was a specific eIF-2 protein kinase finally came from studies of the dsRNA-mediated inhibition of protein synthesis, as described below. The essential experiment was the demonstration that HCR required ATP in order to inhibit the binding of met-tRNA$_f$ to 40 S subunits, and that phosphate was transferred from the γ-position of ATP to the α-subunit of eIF-2 concomitantly with this inhibition. The fact that inhibition of HCR required ATP as cofactor could not have been deduced from studies on protein synthesis in crude lysates, since that assay requires ATP.

2. Characteristics of protein synthesis in crude lysates

The minimal requirements for protein synthesis in crude lysates are added KCl and MgCl$_2$, and an ATP/GTP regeneration system, which can be PEP/PEP kinase [11], creatine phosphate/creatine kinase [10] or fructose 1,6 bisphosphate plus NAD [26]. The lysate contains adequate levels of ATP and GTP and amino acids, and sufficient adenylate kinase and nucleoside diphosphate kinase to convert AMP back to ATP and GDP to GTP. If the lysate is passed over a G50 Sephadex column, additional requirements are ATP, GTP, amino acids and spermidine. In order to obtain prolonged synthesis of protein, haemin or extra GTP (1–2 nM) must be added, and in the case of gel-filtered lysates, dithiothreitol and glucose or a phosphorylated sugar [27] (see Table 1). If these additional components are not present, protein synthesis stops after a few minutes, with the characteristics mentioned below. Apart from the inhibition which sets in if these critical components are *absent*, the *presence* of certain other substances leads to essentially identical symptoms; these are low levels of dsRNA [28] or oxidised glutathione [29]. Certain physical treatments of the lysate, namely incubation at high temperature (40–45°C) [30] and exposure to high hydrostatic pressure (greater than 500 atmospheres) [31] cause similar inhibitions of protein

TABLE 1

Inhibitory condition	Antagonists of inhibition				
	Haemin	GTP	cAMP	Glucose-6-phosphate	dsRNA (1–20 μg/ml)
Haem deficiency	+ + +	+ + +	+ + +	(+)	0
High hydrostatic pressure	0	+ + +	+ + +	NT	NT
Preincubation at 42 °C	0	+ +	+ +	+	NT
Addition of HCR	0	(+)	+ +	0	NT
Addition of dsRNA (1–100 ng/ml)	0	(+)	+ + +	0	+ + +
Addition of GSSG	0	(+)	+	+ + +	0
Lysates from ATP-depleted cells	0	NT	+ +	+	NT
Gel-filtered lysates	0	+	+ +	+ +	0

The cases of lysates from ATP-depleted cells and gel-filtered lysates are complex. For complete relief of inhibition it is often necessary to add glucose-6-phosphate, cAMP and dithiothreitol, and these compounds appear to act synergistically.

Haemin is effective in the range of 10–30 μM: GTP at about 2 mM: cAMP at about 5 mM: Glucose-6-phosphate at about 1 mM. Other purines will substitute for cAMP [40] and other phosphorylated sugars for G-6-P [27].

synthesis. The characteristics of inhibition by this wide variety of environmental changes are broadly similar:

(1) Protein synthesis proceeds at normal rates for a few minutes before it slows, abruptly in most cases, but more gradually in the case of gel-filtered lysates, to a low inhibited rate.

(2) Just before the transition to the inhibited state, 40 S/Met-tRNA$_f$/GTP complexes disappear, although the levels of native 40 S ribosomes, Met-tRNA$_f$ and GTP are completely normal [21,32].

(3) Accompanying the transition, polysomes break down to inactive 80 S ribosomes and nascent chains are lost from the ribosomes. This change does not occur in the presence of elongation inhibitors of protein synthesis.

(4) The synthesis of all proteins, including the synthesis of proteins programmed by added mRNA is inhibited [33–35].

(5) Inhibition can be prevented or reversed by the addition of purified initiation factor eIF-2 [23,36] and certain other less well-characterised proteins [37–39].

(6) High levels (5–10 mM) of cAMP, 2-aminopurine, caffeine, theophylline, adenine and certain other purines can prevent and sometimes reverse the inhibitions [40,41].

(7) High levels of ATP (1–5 mM), tend to potentiate, whereas high levels of GTP tend to alleviate the inhibitions [41,42].

(8) When inhibited lysates are mixed with fresh lysate, the resulting mixture is immediately, though often only transiently inhibited [21].

(9) Inhibitions are accompanied by the conversion of the initiation factor eIF-2 to a phosphorylated form [43]. This conversion does not occur when protein synthesis is not inhibited, nor does it occur in response to other inhibitors of protein synthesis which specifically inhibit initiation, such as NaF, ATA, poly(I), edeine, or pactomycin [44].

These characteristics have been interpreted in terms of the following model: In each of the various conditions mentioned above, a latent inhibitor is activated in the lysate. These inhibitors are either eIF-2 specific protein kinases, or inhibitors of eIF-2 phosphatases which lead to the phosphorylation of eIF-2. Such phosphorylation causes an unspecified impairment of function of eIF-2 such that the rate of formation of 40 S/Met-tRNA$_f$/GTP complexes is greatly reduced, leading to a reduction in the rate of initiation of protein synthesis. Agents which prevent the inhibition may act either by preventing the activation of the inhibitors, or by preventing the action of the inhibitors in some way. In the case of the purines which antagonise inhibition, it is possible that they act by inhibiting protein kinases competitively with respect to the ATP substrate. Some of the protein factors which prevent inhibition may act by increasing the rate of dephosphorylation of eIF-2 rather than by decreasing its rate of phosphorylation, and this may possibly be true in the case of phosphorylated sugars.

On the other hand, two recent reports [38,45] suggest the existence of a non-ribosomal bound protein which prevents the inhibitory effects of added eIF-2 kinases; such an entity has been discovered and rediscovered several times in various contexts as a stimulator of protein synthesis in reticulocyte lysates [46,47]. The factor does not appear to inhibit the phosphorylation of eIF-2 or to enhance the dephosphorylation of eIF-2-PO$_4$; rather it acts either by providing an alternative pathway of initiation, which is unlikely, or by allowing phosphorylated eIF-2 to function normally. The role of this factor in normal initiation is unknown. Further discussion of this is presented below in Section 3.

3. How many inhibitors are there?

It is very clear that haemin prevents the inhibition of protein synthesis by antagonising the activation of latent HCR (proHCR), and that HCR is an eIF-

2 kinase as well as an inhibitor of initiation. It is equally clear that the presence of 1–100 ng/ml of dsRNA in the lysate leads to the activation of a similar inhibitory protein kinase, termed Double-stranded RNA activated Inhibitor (DAI) [44] or dsRNA-induced (dsI) [48]. HCR and DAI are not the same protein kinase, (although they phosphorylate the same sites on eIF-2) by the following criteria:

(1) HCR and DAI elute at different positions from gel-filtration columns.
(2) They have different substrate specificities: while HCR appears to be absolutely specific for eIF-2, DAI can phosphorylate certain histones as well.
(3) They possess different self-phosphorylated components.
(4) Their modes of activation are different; DAI has an absolute requirement for dsRNA and ATP and is unaffected by haemin, whereas HCR activation has no absolute requirements for ATP and is prevented by haemin. NEM activates HCR but does not activate DAI.
(5) High levels of cAMP and other 'rescuing' purines inhibit the activation of DAI more strongly than its action, whereas HCR activation is insensitive to these compounds though its activity is strongly suppressed.
(6) Antibodies which neutralise HCR do not affect DAI [49].

These observations suggest the slightly surprising conclusion that there are at least two quite independent mechanisms for phosphorylating and inactivating reticulocyte eIF-2. How many of the other inhibitory conditions act via these entities?

Experiments done in Cambridge leads us to believe that both high temperature and high hydrostatic pressure activate HCR directly. We are in disagreement with Hardesty's group, who have recently presented evidence that these conditions act by a much more complicated series of reactions [50]. They propose that besides HCR, two proteins exist which can activate HCR by a cascade mechanism in which a heat stable protein is activated either by pressure or elevated temperature, and this then activates a heat labile protein, which in turn activates HCR. Activation of the heat labile protein requires ATP, and its action is blocked by soybean trypsin inhibitor, suggesting that HCR may be activated by partial proteolysis. It is not clear what relationship this postulated activation mechanism has to activation of HCR in a lysate incubated without haemin, where inhibition is reversible, and hence unlikely to involve proteolysis. We were unable to block the inhibitory effects of incubation at 42°C by added soybean trypsin inhibitor, and have been unable to detect the heat-stable inhibitor of protein synthesis. Nonetheless, these observations are interesting and will doubtless receive further study.

The two other inhibitory conditions which might act via HCR activation are the addition of GSSG and the removal of phosphorylated sugars by gel filtrations; these conditions may in fact be the same, since GSSG is reduced to GSH by reactions that consume glucose-6-phosphate. Ernst et al. [51] have presented evidence that addition of GSSG to ribosome-free reticulocyte supernatant activates HCR, an activation that is strongly enhanced by addition of ATP and an ATP generating system. It is also clear that eIF-2 becomes phosphorylated in complete lysates incubated with GSSG [43,52]. However, the pathway by which HCR is activated in these circumstances is unclear, and evidence that inhibition is entirely accounted for by the activation of HCR is at present lacking. We have had some difficulty in detecting 'dominant' inhibitors in extracts of this kind, and I believe that phosphorylated sugars may play some other role in the control of protein synthesis. Studies in Baglioni's laboratory [53,54] have suggested that phosphorylated sugars may directly enhance the binding of Met-tRNA$_f$ to 40 S subunits, although the mechanism of this effect is not clear.

A further possibility for which there is some experimental support is that eIF-2 phosphatase activity may be reduced by some of these conditions [55].

To summarise this confusing area; there is strong evidence for the existence of two distinct eIF-2 kinases which inhibit initiation in reticulocyte lysates, HCR and DAI. It is probably fair to say that there is no need to postulate the existence of yet other eIF-2-specific kinases which are activated by the presence of GSSG or the absence of phosphorylated sugars, but their existence cannot be rigorously excluded at present.

4. Evidence that phosphorylation of eIF-2 leads to a reduction in the rate of initiation

The heart of this control system is widely, if not universally, believed to be the phosphorylation of eIF-2. It is also believed that this modification impairs the function of this initiation factor. What evidence supports this belief?

(1) In the case of HCR, its inhibitory and eIF-2 (α-subunit) kinase activities copurify through all known steps [44,56].
(2) The inhibitory and kinase activities share identical heat-inactivation kinetics [44].
(3) During the activation of HCR and DAI, their kinase and inhibitory activities appear in parallel [44,57].
(4) Antibodies against HCR abolish both its kinase and inhibitory activities [49,58].

(5) Compounds like cAMP, caffeine, theophylline, etc., which antagonise the inhibitory activities of HCR and the activation of DAI, inhibit, or inhibit the appearance of, their kinase activities [44].

(6) Both HCR and DAI require ATP as cofactor, and will not work with AMPPCP or AMPPNP [44].

(7) In crude lysates, inhibition is overcome by the addition of purified eIF-2 [23].

(8) Under certain conditions, partial reactions of purified eIF-2 can be inhibited by the addition of HCR and ATP [59–61].

Clearly, most of this evidence is circumstantial; the only direct evidence (point 8) that phosphorylated eIF-2 is inactive is unfortunately quite controversial. Several groups have had difficulty in demonstrating any alteration in the activity of eIF-2 by phosphorylation [62,63], even though it is easy to show that the reaction catalysed by eIF-2 fails in the presence of HCR and ATP when crude ribosomes are used as the source of the factor and the assay is binding Met-tRNA$_f$ to 40 S subunits. However, it seems to be difficult to form stoichiometric complexes between eIF-2, Met-tRNA$_f$ and GTP at physiological concentrations of these components, and both Gupta's group and Ochoa's laboratory have reported the existence of factors which appear to enhance the mutual affinities of these components [60,61]. They both find that phosphorylated eIF-2 fails to be stimulated by these factors. This makes reasonable sense, even though it is not clear what the role of these factors is in protein synthesis. But the straightforward interpretation is clouded by a report that these kinds of factors may act simply by preventing the binding of eIF-2 to the assay tubes [64]. Since it is presently unclear what is the correspondence between these various 'Co-eIF-2's' (as Gupta calls them) it is exceptionally difficult for an outsider to judge the relative merits of the opposing cases. It would be particularly comforting to have some insight into the mechanism of action of these factors; some progress in this direction has been reported by Gupta, who has raised an antibody against his 'Co-eIF-2A' which inhibits protein synthesis in reticulocyte lysates, suggesting that Co-eIF-2A is an essential part of the protein synthesis machinery [65]. However, it is Co-eIF-2B and-C which confer sensitivity to phosphorylation [61].

The simple test for activity of phosphorylated eIF-2 by testing whether it will restore protein synthesis in an HCR-inhibited lysate is uninterpretable owing to the existence of active eIF-2 phosphatase in the lysate [55].

It has been reported that the three subunits of eIF-2 can be separated from one another in the presence of urea with subsequent retention of activity [66]. The α-subunit, which is phosphorylated by HCR and DAI is apparently the subunit which binds GTP, so that phosphorylation probably interferes in some way with GTP binding, which has been shown to be a prerequisite for Met-tRNA$_f$ binding. An attractive possibility which has probably occurred to everyone working in the field, but for which there is at present no evidence is that phosphorylation prevents the recycling of eIF-2, in the following manner. It is known that GTP is essential for eIF-2 function. It is also known that eIF-2 is released from the ribosome when this GTP is hydrolysed to GDP [67]. It is also known that GDP has a much higher affinity for eIF-2 than does GTP [68], and it is not clear how this GDP is released from eIF-2 and replaced by GTP. It is tempting to postulate the existence of a recycling factor, analogous to the bacterial factor EF.Ts, which could perform this function, and further to suppose that phosphorylation might inhibit the action of this element. It is interesting that several workers have found a factor in the supernatant of lysates which stimulates protein synthesis in lysates incubated with HCR, but which is neither an eIF-2 phosphatase, nor an anti-eIF-2 kinase [38,39,45,47,69]. I wonder whether this might not be the missing recycling factor, which could rescue protein synthesis if present in high enough concentration. It is noteworthy that high concentrations of GTP tend to rescue protein synthesis; although some of this effect can be ascribed to the ability of GTP to antagonise HCR activation, it could be that it also helps displace GDP from eIF-2/GDP complexes.

All in all, this is the most confusing and unsatisfactory aspect of this field. In many ways it reflects our lack of understanding of the physiology of initiation; that is, it is probably that eIF-2 kinases act by inhibiting a step or steps in protein synthesis which have not yet been clearly recognised as such.

5. Sites of phosphorylation of eIF-2

It is not yet firmly established how many sites on eIF-2 become phosphorylated by either HCR or DAI. Measurement of the number of moles of phosphate bound per mole of α-subunit give numbers between 1 and 2. We found [44] there to be two major tryptic peptides containing serine phosphate, but there was undigested material and minor components also present, and we did not attempt to characterise the major peptides further to see if they were related derivatives of the same site. Tuazon and Traugh recently reported [70] isolating four tryptic peptides from the α-subunit phosphorylated by HCR, but again, these might all derive from one site. Isoelectric focussing of phosphorylated eIF-2 has never revealed more than one spot, so that if there are two sites, they must be phosphorylated simultaneously. If there is more than one site, it will be important to establish what if any are the effects of having one or the other or both modified. It is clear, however, that HCR and DAI both phosphorylate the same site(s), despite the fact that they are different enzymes [44].

Although HCR and DAI only phosphorylate the α-subunit of eIF-2, other protein kinases phosphorylate the γ-subunit (55 000 mol. wt.). Reticulocytes contain a very active casein/phosvitin kinase which phosphorylates this component [71]. No alterations in activity of eIF-2 have ever been found as a result of this modification, and casein kinase is not an inhibitor of protein synthesis. However, there is no evidence that eIF-2 is ever phosphorylated in its γ-subunit except when purified factor and kinase are incubated together, or when purified eIF-2 is added to the lysate. Apparently, the γ-subunit is not normally accessible to phosphorylation, probably because it is bound to ribosomes, since its phosphorylation can be suppressed in vitro by the addition of 40 S ribosomes and eIF-3 (R.J. Jackson, unpublished results).

Apart from HCR, DAI and casein kinase, no other protein kinase seems to use eIF-2 as substrate. In particular, eIF-2 is not a substrate for cAMP-dependent protein kinase.

6. eIF-2 phosphatase

Considerably less is known about the enzyme(s) which dephosphorylates the α-subunit of eIF-2 than the enzymes which phosphorylate it. A recent study by Safer and Jagus [55] suggests that the lysate contains a very active phosphatase which rapidly dephosphorylates the α-subunit of added phospho-eIF-2, labelled on both α and γ subunits. The γ-phosphate was completely stable in the lysate, and served as a convenient internal standard for the assay. It is possible that this added eIF-2, whose half life was 20 s might not be accurately representative of eIF-2 phosphate bound to ribosomes. Experiments we did some time ago, using eIF-2 phosphorylated either in situ on ribosomes, or exclusively in the α-subunit showed that dephosphorylation did occur, but much more slowly than Safer and Jagus found. We also found that protein phosphatase-1, a ubiquitous phosphatase found in abundance in all tissues, was capable for removing the α-phosphate, but the rate was rather slow. We were also able to locate an endogenous phosphatase in the lysate after fractionation on Sepharose 6B, where it eluted as quite a small (less than 100 000 mol. wt.) entity, but further attempts at purification were unsuccessful, and nobody else has reported any kind of purification or characterisation of the major enzyme involved. However, Safer and Jagus have begun to study the control of the activity in the crude lysate, and report that it is completely inhibited by aurinetricarboxylic acid, partly inhibited by high concentrations of GDP, and unaffected by haemin, phosphorylated sugars or GSSG. Gel filtration of the lysate reduced the phosphatase activity, but it was apparently impossible to restore the activity. This is clearly an area which needs more work. It would be helpful to have purified active phosphatase, and it would also be helpful to understand whether there is any physiological regulation of this activity. Do the phosphatase inhibitors present in muscle extracts inhibit the activity, for example?

7. The properties and activation of HCR

If it is true that eIF-2 phosphatase(s) is not regulated in any physiologically significant way, and that the extent of phosphorylation of eIF-2 determines the rate of protein synthesis in reticulocyte lysates, it follows that control of eIF-2 kinase activity controls the rate of protein synthesis in these cells. The mechanism of control of HCR activity is not well understood in molecular terms, although there is plenty of phenomenological data and certain classes of hypothesis to account for its properties can be ruled out. It is perhaps helpful to begin by describing the available assays for HCR, since they themselves define, to some extent, a variety of different forms of HCR.

7.1. Inhibition of protein synthesis

HCR was originally defined as an inhibitor of protein synthesis which formed when lysates were incubated for long periods in the absence of haemin [12]. It will inhibit protein synthesis in fresh lysates containing optimal concentrations of haemin. This form of HCR is termed 'irreversible HCR'. The degree of inhibition of protein synthesis is not linearly related to the concentration of HCR in the assay; over a certain range, there is a more or less linear relationship between the logarithm of the dilution of the inhibitor and the extent of protein synthesis, and this allows the definition of a unit of HCR as the dilution required to cause 50% inhibition of protein synthesis under specified conditions. It is of interest that lysates contain a potential yield of 20–30 units of HCR/ml; i.e., 1 volume of lysate incubated at 37°C without haemin for 12 h will inhibit synthesis in 20–30 volumes of fresh lysate by 50%. There is thus a considerable concentration of HCR present in lysate, and activation of a small fraction is sufficient to cause a profound inhibition of protein synthesis. The kinetics of inhibition are complex; high levels of

HCR stop protein synthesis more or less completely after a very short lag, but lower levels inhibit less profoundly after progressively longer lags.

Other forms of HCR can be defined, though less easily quantitated by similar assays. If a lysate is incubated for only 30 min at 30 °C in the absence of haemin and mixed with fresh lysate, the kinetics of incorporation are kinky, that is, there is a lag, followed by a reduction in the rate of protein synthesis, followed by a resumption of protein synthesis at close to the initial rate. This behaviour is interpreted in terms of the activities of a reversible form of HCR in the first incubation which is progressively inactivated during the second [74]. To complicate matters, two forms of 'Reversible HCR' can be recognised. Classical reversible HCR can be inactivated by incubation with haemin, while the other, termed 'Intermediate HCR' is only inactivated by haemin in the presence of fresh lysate [73]. A simple model serves to clarify the relationships between the various forms of HCR [72], shown in Figure 1, which represent either different conformers of a single basic entity, or possibly slightly chemically modified forms of the starting apoinhibitor.

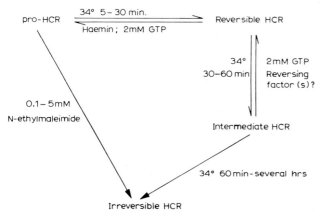

Figure 1. The activation of HCR.

7.2. eIF-2 kinase activity

All forms of HCR, apart from the apoinhibitor, possess specific eIF-2 kinase activity which can be assayed by mixing them with purified eIF-2 and $[\gamma-^{32}P]ATP$ [44]. However, the fact that eIF-2 is a scarce commodity, coupled with the necessity of analysing the incubations on acrylamide gels make this a more difficult and expensive assay than the simple protein synthesis inhibition assay. It is possible to substitute crude ribosomes for purified eIF-2, but they contain very little eIF-2 and a great many other kinases and kinase substrates, giving a very complicated pattern of labelled gel bands, and a rather narrow useful range of sensitivity to HCR. However, kinase assays and protein synthesis inhibitor assays correlate extremely well if carefully performed (see Figure 2).

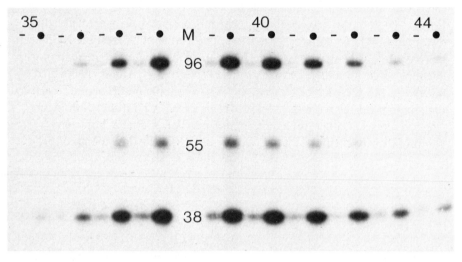

Figure 2. Purification and activation of proHCR. ProHCR was partially purified by chromatography on Sepharose CL-6B and DEAE-Sepharose CL-6B. Fractions containing proHCR were located by incubating portions overnight at 37 °C in the presence of 5 mM dithiothreitol, while the remainder was kept on ice. Each fraction was assayed for its ability to inhibit protein synthesis. No inhibitor was detected in any of the untreated fractions. 8 μl of each fraction, either activated or untreated, were mixed with 1 μg of purified eIF-2 and [γ-^{32}P]ATP at a final concentration of 100 μM and incubated at 30 °C for 10 min, analysed on a gel, and autoradiographed. This photograph shows the autoradiograph. Densitometry of the bands showed a linear correlation between the intensity of the 96 000 mol. wt. self-phosphorylated component of HCR and the 38 000 mol. wt. subunit of eIF-2, and a 10-fold increase in the intensities of both bands in the activated samples of each fraction. Figures along the top indicate fraction numbers; M stands for marker track; —— fractions kept on ice; $\bullet$, fractions activated by incubation at 37 °C. Notice that some activity against the largest (55 000 mol. wt.) subunit of eIF-2 is induced, but that the intensity of this band is only a tenth of that of the α-subunit. In the absence of added initiation factor, one would only see the labelling of the 96 000 mol. wt. band. The staining pattern failed to show any protein band corresponding to the 96 000 mol. wt. band of radioactivity.

7.3. Self-phosphorylation of HCR

An alternative to the eIF-2 kinase assay is the autokinase activity of HCR itself. When preparations of HCR are incubated briefly with $[\gamma\text{-}^{32}P]ATP$, and the incubations analysed on acrylamide gels, a labelled band migrating between 80 and 100 000 mol. wt. is found (the precise molecular weight of this band will be discussed below) whose intensity is correlated with HCR activity as measured by either of the other two assays [75]. Since this assay is one of extent, rather than rate of labelling as with the eIF-2 kinase assay, it is much easier to analyse a large number of samples (from a column, say) at once. The assay is also very tolerant of a wide variation in salt and pH, and is extremely sensitive if high specific activity ATP is used. If is often possible to detect HCR in column effluents by this assay when for various reasons like dilution or the presence of high salt concentrations, it is impossible to detect by the standard protein synthesis test.

7.4. Assay for proHCR

The assay and purification of proHCR is complicated by its lack of activity in any of the foregoing assays. True proHCR has no activity at all until after its activation, either by incubation for some time at 40 °C or by adding N-ethylmaleimide to a final concentration of 0.5 mM in excess of any thiols present and incubation for 30 min at 30 °C [78]. The HCR thus activated can then be assayed as described above.

8. Purification of HCR

HCR has been purified by Gross and Rabinovitz [17] and by Ranu and London [76] using standard protein purification procedures. The purification of reversible HCR has also been reported [77], but to date, proHCR has not been completely purified. It is generally found that the yields are poor, 5% seems to be a typical figure, which might raise questions as to whether the purified material is fully representative of the starting HCR activity. It is universally agreed that HCR does not change its size, charge or shape on activation, which appears to rule out models for activation such as the $R_2C_2 \rightleftharpoons R_2 + C_2$ generally accepted for the cAMP-dependent protein kinases, and may also rule out phosphorylation as a necessary prerequisite for activation; but more of that anon.

9. The size, shape and subunit composition of HCR

There are wide variations in the literature describing the molecular parameters
of HCR. Values for the molecular weight range from 60 000 to 500 000,
depending on the preparation of HCR and the method used to determine it
[76–79]. There is in fact a way to reconcile the various estimates, for those
who have used gel filtration have obtained high values, of the order of 320 000
[78,79], while those who have used sucrose gradient centrifugation obtain
much lower values, of the order of 100 000 or less [76,77]. Yet the autokinase
which seems to be an integral part of the enzyme has a molecular weight on
SDS-acrylamide gels of 96 000 [44,75]. I have confirmed all these observations
with the same partially pure preparation of HCR run in the same buffer on all
systems. If one combines the data from the stokes radius obtained by gel
filtration with the S value obtained by sucrose density gradient centrifugation,
the method of Siegel and Monty [80] yields a value of 160 000 for the
molecular weight, with a frictional ratio of 1.73. In order to determine the
subunit composition, and also to check that the low S value did not result
from pressure-induced dissociation on the gradient, HCR was labelled with
$[\gamma^{-32}P]ATP$ and the labelled protein cross-linked with dimethylsuberimidate.
A single new band appeared on SDS acrylamide gels with exactly twice the
apparent molecular weight of the labelled monomer. This cross-linked species
sedimented in exactly the same position as the unmodified protein. I therefore
believe that HCR is a homodimer with two subunits of 80 000 and that it has
an elongated shape. It seems that the labelled autokinase band runs
anomalously on gels, and depending on the precise formulation of the gel, in
particular the percentage gel and the proportion of bisacrylamide used, values
between 80 000 and 105 000 can be obtained. In agreement with others, I have
been unable to detect changes in sedimentation velocity or elution position
from gel-filtration columns between proHCR and irreversible HCR. It does
not seem that any major change in configuration accompanies activation,
which is perhaps disappointing, and may or may not be significant. Nor is
there any evidence that purified HCR has different physical properties from
crude HCR.

10. The activation of HCR

Several models for the activation of HCR have been proposed, and the matter
cannot be regarded as settled with present information. I will begin by
reviewing the various ways in which proHCR can be activated.

10.1. Incubation in the absence of haemin

Whether crude lysates or highly purified proHCR preparations are taken, incubation at physiological temperature in the absence of haemin progressively activates HCR, probably by the series of intermediates described by Gross [72], that is, via reversible and intermediate HCR to give irreversible HCR. There is no obvious requirement for any low molecular weight cofactors in this process, and in particular it should be stressed perhaps that ATP is not necessary, although in crude lysates it does appear to enhance the formation of intermediate HCR [42]. The presence of haemin retards this activation very markedly, except when the temperature is raised above about 40 °C. The rate of activation increases with increasing temperature, and proceeds very slowly below 25 °C, often showing a lag at low temperatures. It appears to be independent of the dilution of the proHCR, whether crude or purified, implying that the rate limiting step in activation is unimolecular. It certainly appears reasonable to interpret these data to indicate (as originally proposed by Rabinovitz) that proHCR is inherently unstable and undergoes a spontaneous change in conformation unless haemin is present to stabilise it in its inactive state.

A curious feature of proHCR is that high concentrations (2 mM) of GTP prevent its activation, and are highly effective at converting intermediate HCR back to proHCR, more so than haemin. ATP competes with GTP in this reaction [42]. It is not clear whether GTP hydrolysis is required. This aspect of HCR behaviour is poorly understood, both as regards mechanism and physiological significance. I shall return to this question later.

10.2. Activation by −SH reactive reagents

When crude lysates are incubated with 5–10 mM N-ethylmaleimide, HCR is rapidly activated regardless of whether haemin is present or not [82]. The HCR so formed seems to be irreversible HCR, and does not appear to be distinguishable from HCR activated by prolonged incubation in the absence of haemin. Other −SH reactive reagents, such as α-iodosobenzoate act similarly. At high temperatures, HCR is inactivated by NEM, and higher concentrations of NEM (i.e., 10 mM) which activate the HCR in crude lysates inactivate the HCR in gel filtered lysates or purified proHCR. Investigation of this curious effect showed that the component removed by gel filtration which protected against inactivation and allowed activation was ATP. Lower concentrations of NEM (0.5 mM) will still activate purified proHCR in the absence of ATP [81]. These data suggest that HCR possesses two distinct sites for reaction with NEM, one of which leads to activation, the other which leads to inactivation. Curiously, the inactivation site appears to be more accessible to NEM in proHCR than in activated HCR, although it requires a higher concentration of NEM to react. At 40 °C, HCR is rapidly inactivated by NEM.

There are other curious features about this activation/inactivation. When partially purified proHCR is reacted with 10 mM NEM, it seems to be irreversibly inactivated, and never shows any sign of activity. Yet if the same preparation of proHCR is first incubated with 0.5 mM NEM until fully active, and then the concentration of NEM raised to 10 mM, subsequent activation is rather slow, unless the temperature is raised to 40 °C. Thus there is a paradox: HCR is activated by low concentrations of NEM, inactivated by high concentration. But activated HCR is much more slowly inactivated by high concentrations of NEM than proHCR. I interpret these observations to indicate that the active site (which reacts with high concentrations of NEM, and can be protected by ATP) is more accessible to NEM in proHCR than in active HCR; this seems to provide clear evidence of a conformational change, as reflected in the accessibility of a presumed cysteine residue to NEM, upon activation. It would be most interesting to confirm these conjectures by a study of the peptides labelled with [^{14}C]NEM under various circumstances.

The activation of proHCR by NEM suggests that protein $-$SH groups may be involved in the activation process, in which case, thiol protective agents might tend to prevent activation. Gross has reported [83] that this is indeed the case, but I have been unable to confirm this; indeed, to the contrary, high levels (5–10 mM) of dithiotreitol tend to promote activation of crude proHCR. This is particularly vividly illustrated by some attempts to purify proHCR on thiopropyl Sepharose 6B. ProHCR binds quantitatively to this resin, in the 2-thiopyridine derivatised form. It can be eluted at pH 8 with DTT, but comes off fully activated! The fact that it can be eluted suggests that its $-$SH groups must be fully reduced, and the fact that NEM-activated HCR does not bind to the resin suggests that binding does occur by covalent attachment through S–S bridges. It may be that any substitution of the critical $-$SH group leads to irreversible activation of HCR, and that removal of the substituent does not necessarily lead to restoration of the previous conformation.

10.3. Activation of proHCR at high pH

Further hints that $-$SH groups might be intimately involved in the activation of proHCR came from studies on the effect of pH on activation [81]. Partially purified proHCR was incubated in PIPES, HEPES or Tris buffers at a variety of pHs. Overlapping ranges of the different buffers were used to check that they had no effect on activation other than by the pH of their solutions. Samples were incubated at 30 °C in the absence of haemin for 40 min, and assayed for their HCR content by their capacity to inhibit protein synthesis. At high pH, HCR was activated, and the apparent pK of activation was about pH 8.4, close to the pK of cysteine. At high pH (9.5) activation was virtually complete and instantaneous even at 0 °C, whereas at pH 6.5 there was very little activation at all. This observation is of some practical importance if one is trying to purify proHCR – one should keep the pH of the buffers low.

10.4. Activation of proHCR at high hydrostatic pressure

When lysates are exposed to high hydrostatic pressure at 0 °C in the presence of optimal concentrations of haemin, they rapidly generate an inhibitor of protein synthesis which resembles intermediate HCR in its properties, that is, it is not inactivated by incubation with haemin, but is inactivated by incubation with 2 mM GTP. It elutes from gel filtration columns in the same place as classically activated HCR, and the inhibitory fractions show increased autokinase activity. I have been unable to detect the low molecular weight inhibitor described by Henderson and Hardesty [31] as being formed under these circumstances, despite several attempts, including analysis of samples sent by them. I do not know why our results differ; there must be some differences in our assay conditions.

One might suppose that this activation indicates that activated HCR has a lower volume that proHCR, but other interpretations are no doubt possible. A recent report claims that high pressure causes partial oxidation of –SH groups in lactate dehydrogenase [84]; might this be so in the case of proHCR? But if this is so, why should GTP be so effective at inactivating the inhibitor?

10.5. The role of cAMP in the activation of HCR

It has been known for some time that beef heart cAMP-dependent protein kinase will inhibit protein synthesis in the lysate if added at high concentrations in the presence of 50–100 μM cAMP [85,86]. The inhibition caused by these additions takes quite a long time to set in, and is rather gradual compared to the inhibition caused by lack of haemin or by dsRNA. The reason for this inhibition is unknown, but the fact that cAMP is necessary for the beef heart enzyme to inhibit does strongly suggest that protein phosphorylation is involved. Since eIF-2 is not a substrate for this enzyme, if the inhibition is due to phosphorylation of eIF-2, it follows that either HCR or DAI are being activated, or that eIF-2 phosphatase is being inactivated in these circumstances. Ochoa and his colleagues proposed [86,87] that this inhibition was due to activation of HCR by phosphorylation, and went further by suggesting that the normal activation of HCR in the absence of haemin was due to an endogenous cAMP-dependent protein kinase whose activity was inhibited by haemin [88]. They showed that haemin could inhibit the cAMP-induced dissociation and activation of purified cAMP-dependent protein kinase, that it bound to the regulatory subunit of this enzyme, and inhibited the binding of cAMP thereby. This model has considerable appeal, particularly since ATP appears to promote the formation of intermediate HCR, but it rests on shaky arguments, and there is rather compelling evidence that it is incorrect.

First, it might be expected that cAMP itself would inhibit protein synthesis in the lysate at certain concentrations, but such is not the case, even though it can be shown that the presence of cAMP leads to enhanced phosphorylation of several proteins in the lysate. This phosphorylation is not diminished by the presence of haemin [81,89]. Therefore it seems that physiological concentrations of haemin do not inhibit cAMP-dependent protein kinase in the crude lysate; either the endogenous reticulocyte enzyme is not affected by haemin, or the concentration of free haemin in the lysate is insufficient to inhibit the enzyme. Second, if this model is correct, inhibition of the cAMP-dependent protein kinase ought to make a lysate insensitive to haemin. It is possible to inhibit cAMP-dependent protein kinase with the heat-stable kinase inhibitor from muscle. Addition of this inhibitor to lysates incubated without added haemin does not stimulate protein synthesis in the way the model predicts [81,89,90]. It can be shown, however, that it does inhibit the cAMP-enhanced phosphorylation of reticulocyte proteins, and the activity of added cAMP-dependent protein kinase in the lysate [81]. Third, the rate of activation of partially purified proHCR is not increased by addition of cAMP-dependent protein kinase, cAMP and ATP. Fourth, purified catalytic subunit of the rabbit muscle cAMP-dependent protein kinase does not inhibit protein synthesis in the lysate [90]. Given these data, which have been obtained in at least three laboratories, it seems most likely that the results reported by Ochoa's group were due to contaminating activities in the kinase preparations they used. The available data, taken as a whole, do not support models which propose enzymic modification of proHCR as a basis for its activation. It seems more likely that a conformational change in its structure occurs, the probability of this change occurring being influenced by the presence of ligands like haemin, ATP and GTP. It is unclear whether modification of

−SH groups is a physiological occurrence. It could be that oxidation by superoxide or hydroxyl radicals occurs, and that haemin or GTP make the sensitive −SH group less accessible. However, the fact that high concentrations of mercaptoethanol and DTT tend, if anything, to promote the activation of proHCR rather than prevent it, and the lack of reproducible effects of H_2O_2 or CN^- (which inhibits catalase) offer no support to this idea. However, it seems that free radical reactions are poorly understood in these complex living systems, and I would not be dogmatic. In concluding this section I would like to point out two things. First, the affinity of proHCR for haemin must be rather low compared to globin and even human serum albumin, which also forms stoichiometric complexes with haemin, otherwise the control mechanism would not work. Second, I still find it hard to accept that it seems to be the absence of something which turns on proHCR. One is conditioned to the idea that control mechanisms work in a more positive kind ·of way, that the accumulation of something rather than the depletion of something is likely to act as the switch, although there is no good reason why this should be so. Deoxyhaemoglobin has a different conformation from oxyhaemoglobin, and the change is completely and autogenously reversible. The lac repressor inhibits lac mRNA synthesis in the absence of an inducer, and permits it in the presence of the inducer. Again, significant conformational changes leading to profound alterations in the properties occur fully and reversibly without any covalent bond making or breaking. In the case of HCR, it is perfectly reasonable that mild denaturation caused by warming or modification of −SH groups should lock the structure in the active, rather than the inactive conformation.

11. Self-phosphorylation of HCR

Having said all this, it is important to remember that active HCR is apparently phosphorylated, while proHCR is not. We noticed very early on [44] that different preparations of HCR always contained a phosphorylated component which runs with an apparent molecular weight of 96 000 on our standard gel system, although it probably has a molecular weight of 80 000, as already discussed. The intensity of this labelled component was closely ·correlated with the activity of HCR whether assayed as an inhibitor of protein synthesis or as eIF-2 kinase. The labelling of this component is very slight in preparations of proHCR, and can be fully supressed by haemin in these preparations, although haemin has no effect on the labelling of activated HCR [77]. Antibodies against HCR also inhibit the labelling of this component [49]. Kinetic studies of the labelling show that it is an intramolecular reaction, for it is very rapid and unaffected by the concentration of HCR over a wide range [75,81]. There is every reason to suppose that this self-phosphorylation represents an intrinsic activity of HCR. Labelling occurs exclusively on serine residues, but the number of phosphates per mole of 80 000 subunit remains to be determined. Gross estimated that there might be as many as 5 phosphates per mole, based on the amount of stained HCR he detected on a gel [75]. I have tried to determine this stoichiometry by phosphopeptide analysis, which is a difficult task. There are certainly more than one phosphopeptide produced by digestion with a variety of enzymes, and cyanogen bromide treatment of labelled HCR yielded two bands on an acrylamide gel and very little, if any, loss of label. One of the bands had exactly three times the radioactivity of the other, suggesting a minimum number of four phosphates (it could in principle be a multiple of four) per subunit). It has recently been shown that HCR becomes labelled during incubation in the absence of haemin in the lysate [52], but it is not clear whether it loses the phosphate when haemin is added back. I found that the phosphate label on irreversible HCR was completely stable when it was incubated with lysate under a variety of conditions. Because it is difficult to answer the question as to whether or not the phosphate is removed from HCR when it is inactivated under physiological conditions, which might shed some light on the role of this phosphorylation, because it is very hard to resolve labelled HCR from other phosphoproteins in the lysate. I have been unable to use any two-dimensional gel analysis since HCR seems to streak and not run as a tight spot in any system other than the standard SDS-acrylamide gel.

Comparison of the inhibitory activity of HCR with its autokinase activity suggest that one unit of inhibitor contains about 10 pmol of phosphate, which in turn means that one unit of inhibitor is about 0.2 μg, and that an average lysate contains about 5 μg/ml of HCR, all assuming that there are 4 phosphates per 80 000 mol. wt. subunit. In vitro, one unit of HCR can phosphorylate about 20 pmol of eIF-2 per minute, which corresponds reasonably well with the estimates that lysate contains about 20 pmol/ml of eIF-2.

12. The dsRNA activated protein kinase/inhibitor

The response of the reticulocyte lysate to added dsRNA is essentially identical to that of cell-free systems prepared from cells that have been exposed to physiological doses of interferon. It is completely unknown how and when reticulocytes acquire, or why they possess this regulatory system.

Protein synthesis in the lysate is exquisitely sensitive to inhibition by dsRNA [91]. Inhibition can be caused by as little as 0.1 ng/ml of viral dsRNA, which is only 0.15 nM in base pairs, or about 0.03 pM in molecules. Put another way, this is roughly the equivalent of one molecule of, say, poliovirus dsRNA in one HeLa cell. Clearly, dsRNA must inhibit protein synthesis catalytically, given that the lysate is about 250 nM in ribosomes and roughly 50 nM with respect to globin mRNA. Evidence that this is indeed so comes from experiments in which lysate is preincubated with dsRNA, and then mixed with fresh lysate. It is found that the preincubation generates an inhibitor of initiation which resembles HCR in many respects, although as pointed out earlier, is not HCR. Formation of this inhibitor, which is an eIF-2 α-subunit kinase, requires ATP and non-hydrolysable analogues of ATP and other nucleotide triphosphates fail to substitute for ATP [44]. Accompanying the activation of this inhibitor, a protein with a molecular weight of 67 000 on SDS acrylamide gels becomes phosphorylated, and the best evidence, though not conclusive, implies that this protein is a component of the inhibitor [92]. At present, the inhibitor has not been purified from reticulocytes, and its structure and mode of activation are unknown. One of the most curious features of this system (apart from the puzzling question as to why it is present in reticulocytes at all) is that high concentrations of dsRNA, above about 1 μg/ml fail to inhibit protein synthesis, and do not activate the eIF-2 kinase [91]. In some lysates, addition of a 'high' dose of dsRNA after a low, inhibitory dose can actually restore protein synthesis to its initial rate. One can construct models which would account for this paradoxical behaviour, such as postulate that there is a high dsRNA activated phosphatase which switches off the DAI, or that DAI has two components which must bind adjacently on dsRNA, but as far as I know, there is no information presently available which might explain the effect.

To complicate matters, reticulocyte lysates have been shown to synthesise the low molecular weight inhibitor of protein synthesis pp5'A2'p5'A2'p5'A (2'5' A) in the presence of dsRNA [93]. This compound inhibits protein synthesis by activating an endogenous nuclease which degrades mRNA [94]. Given the existence of two independent inhibitory systems which are both activated by dsRNA, one may ask which is the predominant one? The answer depends to a large extent on the conditions of incubation of the lysate [95]. Under the ionic conditions we normally use, the kinase system is responsible for the observed inhibition of protein synthesis, but it is possible, by using poly I:C rather than natural dsRNA, and raising the salt concentration, to obtain inhibition of protein synthesis almost exclusively via the 2'5'A mediated pathway. It should be emphasized that 2'5'A has no direct effects on the initiation of protein synthesis and does not appear to activate inhibitory protein kinases.

13. Control of translation in other cell-free systems

Reticulocytes are not the only cells to show clear evidence of possessing a control mechanism which regulates the rate of initiation of protein synthesis. Most cell types studied show superficially similar symptoms (a reduction in the rate of protein synthesis and a concomitant conversion of polysomes to inactive 80 S ribosomes) when subjected to a variety of environmental stresses. Alterations in the temperature or the tonicity of the medium or starvation for essential nutrients such as growth factors, glucose or amino acids cause such effects [96]. Unfortunately, no other cell-free system shows the prolonged activity of the reticulocyte lysate, even when haemin, GTP or high levels of cAMP are added, although a reduction in the rate of initiation seems to account for the effect. We have been unable to detect any phosphorylation of eIF-2 during incubations of these systems, and significantly, protein synthesis is not restored by addition of this initiation factor. On the other hand, protein synthesis can be inhibited by addition of HCR, and under these circumstances, eIF-2 is phosphorylated. Conversely, we have been unable to detect HCR in these cell extracts after they stop making protein, and there is rather little evidence that they stop making protein because they accumulate an inhibitor of initiation like HCR or DAI. In order to inhibit reticulocyte lysate protein synthesis it is necessary to add quite large amounts of the other cell extract (1:1 mixtures). Cimadevilla et al. [97] have briefly described an inhibitor of translation from Friend leukaemia cells, which does not appear to be HCR, and whose mechanism of action is still unknown.

Inhibitors of protein synthesis which seem to resemble HCR have been isolated from Ehrlich ascites tumour cells [98] and rat liver [99], but they have not been thoroughly characterised. In particular, the extracts were not deliberately activated in any way, and the concentration of the HCR-like activity was low compared to reticulocytes. There is some suggestive evidence that inhibitors of this kind may regulate protein synthesis in various cell types from experiments reported by Baglioni and his colleagues. They found that haemin could stimulate protein synthesis to some extent when added to the homogenisation buffer of HeLa cells [100], curiously, addition of haemin to the clarified homogenate failed to stimulate protein synthesis. It is unclear whether haemin stimulates these systems by preventing the activation of HCR or related enzymes. There is no evidence that iron deficiency leads to inhibition of protein synthesis in HeLa cells. Baglioni's group have also reported [53] that phosphorylated sugars stimulate protein synthesis in extracts of mammalian tumour cells, and draw parallels between this effect and the stimulation of protein synthesis in sugar phosphate depleted reticulocyte lysates; but the exact way in which these compounds act remains obscure. There is very little firm evidence that they prevent the activation of translational inhibitors, indeed, West et al. favour the idea that they directly promote the binding of Met-tRNA$_f$ to native 40 S subunits [54].

There are indications that other control mechanisms may operate in these systems, which may or may not involve phosphorylation of components of the protein synthesis machinery. For example, it has long been known that ribosomal protein S6 can exist in a variety of phosphorylated forms, and despite great uncertainty as to whether changes in the phosphorylation state of this protein correlate with changes in the activity of ribosomes, two recent reports tend to support the idea that inactive ribosomes have very little phosphorylated S6 [101,102]. It may be that S6 has to be phosphorylated for ribosomes to be active, and that in vitro, phosphatases are more active than the (unidentified) kinases; it is considerably more difficult to analyse a situation in which dephosphorylation causes inhibition of protein synthesis than one in which phosphorylation leads to inhibition, since the simple experiments using [γ-^{32}P]ATP which are revealing in the latter case are useless in the former. My personal feeling is that phosphorylation of eIF-2 is not the only way in which protein synthesis is regulated in mammalian cells, indeed, that this particular kind of control mechanism may be restricted to reticulocytes and interferon-treated cells. There is an urgent need to make detailed studies of cell-free systems from other cell types to find out why protein synthesis stops in vitro, and how it can be prolonged.

References

1. Kruh, J. and Borsook, H. (1956) J. Biol. Chem., 220, 905–915.
2. Borsook, H., Fischer, E.H. and Keighley, G. (1957) J. Biol. Chem., 229, 1059–1070.
3. Bruns, G.P. and London, I.M. (1965) Biochem. Biophys. Res. Commun., 18, 236–242.
4. Waxman, H.S. and Rabinovitz, M. (1965) Biochim. Biophys. Acta, 129, 369–379.
5. Grayzel, A.I., Hörchner, P. and London, I.M. (1966) Proc. Natl. Acad. Sci. USA, 55, 650–655.
6. Karibian, D. and London, I.M. (1965) Biochem. Biophys. Res. Commun., 18, 243–249.
7. Lamfrom, H. and Knopf, P.M. (1964) J. Mol. Biol., 9, 558–575.
8. Zucker, W.V. and Schulman, H.M. (1967) Biochim. Biophys. Acta, 138, 400–410.
9. Hammel, C.L. and Bessman, S.P. (1966) Science, 152, 1080–1082.
10. Zucke, W.V. and Schulman, H.M. (1968) Proc. Natl. Acad. Sci. USA, 59, 582–589.

11. Adamson, S.D., Godischaux, W. and Herbert, E. (1968) Arch. Biochem. Biophys., 125, 671–683.
12. Maxwell, C.R. and Rabinovitz, M. (1969) Biochem. Biophys. Res. Commun., 35, 79–85.
13. Rabinovitz, M. Freedman, M.L., Fischer, J.M. and Maxwell, C.R. (1969) Cold Spring Harbor Symp. Quant. Biol., 34, 567–578.
14. Burka, E.R. (1968) Science, 162, 1287.
15. Felicetti, L., Colombo, B. and Baglioni, C. (1966) Biochim. Biophys. Acta, 129, 380–394.
16. Waxman, H.S., Freedman, M.L. and Rabinovitz, M. (1967) Biochim. Biophys. Acta, 145, 353–360.
17. Gross, M. and Rabinovitz, M. (1973) Biochem. Biophys. Res. Commun., 50, 832–838.
18. Darnbrough, C., Legon, S., Hunt, T. and Jackson, R.J. (1973) J. Mol. Biol., 76, 378–403.
19. Schreier, M.H. and Staehelin, T. (1973) Nature New Biol., 242, 35–38.
20. Levin, D.H., Kyner, D. and Acs, A. (1973) J. Biol. Chem., 248, 6416–6425.
21. Legon, S., Jackson, R.J. and Hunt, T. (1973) Nature New Biol., 241, 150–152.
22. Benne, R., Wong, C., Luedi, M. and Hershey, J.W.B. (1976) J. Biol. Chem., 251, 7675–7681.
23. Clemens, M.J., Henshaw, E.C., Rahamimoff, H. and London, I.M. (1974) Proc. Natl. Acad. Sci. USA, 2946–2950.
24. Balkow, K., Mizuno, S. and Rabinovitz, M. (1975) Biochem. Biophys. Res. Commun., 54, 315–323.
25. Gross, M. (1979) J. Biol. Chem., 254, 2370–2383.
26. Jackson, R.J. and Hunt, T. (1978) FEBS Lett., 93, 235–238.
27. Ernst, V., Levin, D.H. and London, I.M. (1978) J. Biol. Chem., 253, 7163–7172.
28. Ehrenfeld, E. and Hunt, T. (1971) Proc. Natl. Acad. Sci. USA, 68, 1075–1078.
29. Kosower, N.S., Vanderhoff, G.A., Benerofe, B., Hunt, T. and Kosower, E.M. (1971) Biochem. Biophys. Res. Commun., 45, 816–821.
30. Bonanou, Tzedaki, S.A., Smith, K.E., Sheeran, B.A. and Arnstein, H.R.V. (1978) Eur. J. Biochem., 84, 601–610.
31. Henderson, A.B. and Hardesty, B. (1978) Biochem. Biophys. Res. Commun., 83, 715–723.
32. Darnbrough, C.H., Hunt, T. and Jackson, R.J. (1972) Biochem. Biophys. Res. Commun., 48, 1556–1564.
33. Mathews, M.B., Hunt, T. and Brayley, A. (1973) Nature New Biol. 243, 230–233.'
34. Beuzard, Y., Rodvien, R. and London, I.M. (1973) Proc. Natl. Acad. Sci. USA, 70, 1022–1026.
35. Lodish, H.F. and Desalu, O. (1973) J. Biol. Chem., 248, 3520–3527.
36. Kaempfer, R.O. (1975) Biochem. Biophys. Res. Commun., 61, 591–597.
37. Ralston, R.O., Das, A., Dasgupta, A., Roy, R., Palmieri, S. and Gupta, N.K. (1978) Proc. Natl. Acad. Sci. USA, 75, 4858–4862.
38. Amesz, H., Goumans, H., Haubrich-Morree, T., Voorma, H.O. and Benne, R. (1979) Eur. J. Biochem., 98, 513–520.
39. Ralston, R.O., Das, A., Dasgupta, A., Roy, R., Palmieri, S. and Gupta, N.K. (1978) Proc. Natl. Acad. Sci. USA, 75, 4858–4862.
40. Legon, S., Brayley, A., Hunt, T. and Jackson, R.J. (1974) Biochem. Biophys. Res. Commun., 56, 745–752.
41. Ernst, V., Levin, D.H., Ranu, R.S. and London, I.M. (1976) Proc. Natl. Acad. Sci. USA, 1112–1116.
42. Balkow, K., Hunt, T. and Jackson, R.J. (1975) Biochem. Biophys. Res. Commun., 67, 366–375.
43. Farrell, P.J., Hunt, T. and Jackson, R.J. (1978) Eur. J. Biochem., 89, 517–521.
44. Farrell, P.J., Balkow, K., Hunt, T., Jackson, R.J. and Trachsel, H. (1977) Cell, 11, 187–200.
45. Ralston, R.O., Das, A., Grace, M., Das, H. and Gupta, N.K. (1979) Proc. Natl. Acad. Sci. USA, 76, 5490–5494.

46. Kosower, N.S., Vanderhoff, G.A. and Kosomer, E.M. (1972) Biochim. Biophys. Acta, 272, 623–637.
47. Gross, M. (1976) Biochim. Biophys. Acta, 447, 445–459.
48. Levin, D. and London, I.M. (1978) Proc. Natl. Acad. Sci. USA, 75, 1121–1125.
49. Petryshyn, R., Trachsel, H. and London, I.M. (1979) Proc. Natl. Acad. Sci. USA, 76, 1575–1579.
50. Henderson, A.B., Miller, A.H. and Hardesty, B. (1979) Proc. Natl. Acad. Sci. USA, 76, 2605–2609.
51. Ernst. V., Levin, D.H. and London, I.M. (1978) Proc. Natl. Acad. Sci. USA, 75, 4110–4114.
52. Ernst, V., Levin, D.H. and London, I.M. (1979) Proc. Natl. Acad. Sci. USA, 76, 2118–2122.
53. Lenz, J.R., Chatterjee, G.E., Moroney, P.A. and Baglioni, C. (1978) Biochemistry, 17, 80–87.
54. West, D.K., Lenz, J.R. and Baglioni, C. (1979) Biochemistry, 18, 624–632.
55. Safer, B. and Jagus, R. (1979) Proc. Natl. Acad. Sci. USA, 76, 1094–1098.
56. Trachsel, H., Ranu, R.S. and London, I.M. (1978) Proc. Natl. Acad. Sci. USA, 75, 3654–3658.
57. Gross, M. and Mendelewski, J. (1977) Biochem. Biophys. Res. Commun., 74, 559–569.
58. Kramer, G., Cimadevilla, J.M. and Hardesty, B. (1976) Proc. Natl. Acad. Sci. USA, 73, 3078–3082.
59. Kramer, G., Henderson, A.B., Pinphanichakarn, P., Wallis, M.A. and Hardesty, B. (1977) Proc. Natl. Acad. Sci. USA, 74, 1445–1449.
60. de Haro, C. and Ochoa, S. (1978) Proc. Natl. Acad. Sci. USA, 75, 2713–2716.
61. Das, A., Ralston, R.O., Grace, M., Roy, R., Ghosh-Dastidar, P., Das, H.K., Yaghmai, B., Palmieri, S. and Gupta, N.K. (1979) Proc. Natl. Acad. Sci. USA, 76, 5076–5079.
62. Trachsel, H. and Staehelin, T. (1978) Proc. Natl. Acad. Sci. USA, 75, 204–208.
63. Safer, B. and Anderson, W.F. (1978) Crit. Revs. Biochem., 6, 261–289.
64. Benne, R., Amesz, H., Hershey, J.W.B. and Voorma, H.O. (1979) J. Biol. Chem., 254, 3201–3205.
65. Ghosh-Dastidar, P., Yaghmai, B., Das, H.K. and Gupa, N.K., personal communication.
66. Barrieux, A. and Rosenfeld, M.G. (1977) J. Biol. Chem., 252, 392–398.
67. Peterson, D.T., Safer, B. and Merrick, W.C. (1979) J. Biol. Chem., 254, 7730–7735.
68. Walton, G.M. and Gill, G.N. (1975) Biochim. Biophys. Acta, 390, 231–245.
69. Gross, M. (1975) Biochem. Biophys. Res. Commun., 67, 1507–1515.
70. Tuazon, P.T. and Traugh, J.A. (1979) Fed. Proc. 38, 781.
71. Tahara, S.M., Traugh, J.A., Sharp, S.B. Inndak, T.S., Safer, B. and Merrick, W.C. (1978) Proc. Natl. Acad. Sci. USA, 75, 789–793.
72. Gross, M. (1974) Biochim. Biophys. Acta, 340, 484–497.
73. Gross, M. (1974) Biochim. Biophys. Acta, 366, 319–332.
74. Gross, M. and Rabinovitz, M. (1972) Proc. Natl. Acad. Sci., 69, 1565–1568.
75. Gross, M. and Mendelewski, J. (1978) Biochim. Biophys. Acta, 520, 650–663.
76. Ranu, R.S. and London, I.M. (1976) Proc. Natl. Acad. Sci. USA, 73, 4349–4353.
77. Trachsel, H., Ranu, R.S. and London, I.M. (1978) Proc. Natl. Acad. Sci. USA, 75, 3654–3658.
78. Gross, M. and Rabinovitz, M. (1972) Biochim. Biophys. Acta, 287, 340–352.
79. Adamson, S.D., Yau, P.M.-P., Herbert E. and Zucker, W.V. (1972) J. Mol. Biol., 63, 247–264.
80. Siegel, L.M. and Monty, K.J. (1966) Biochim. Biophys. Acta, 112, 346–362.
81. Hunt, T. (1979) Miami Winter Symp., 16, 321–345.
82. Gross, M. and Rabinovitz, M. (1972) Biochim. Biophys. Acta, 287, 340–352.
83. Gross, M. (1978) Biochim. Biophys. Acta, 520, 642–649.
84. Schmid, G., Ludemann, H.-D. and Jaenicke, R. (1978) Eur. J. Biochem., 86, 219–224.

85. Ernst, V., Levin, D.H., Ranu, R.S. and London, I.M. (1976) Proc. Natl. Acad. Sci. USA, 73, 1112–1116.
86. Datta, A., deHaro, C., Sierra, J.M. and Ochoa, S. (1977) Proc. Natl. Acad. Sci. USA, 74, 1463–1467.
87. Datta, A., deHaro, C., Sierra, J.M. and Ochoa, S. (1977) Proc. Natl. Acad. Sci. USA, 74, 3326–3329.
88. Datta, A., deHaro, C. and Ochoa, S. (1978) Proc. Natl. Acad. Sci. USA, 75, 1148–1152.
89. Levin, D.H., Ernst, V. and London, I.M. (1979) J. Biol. Chem., 254, 7935–7941.
90. Grankowski, N., Kramer, G. and Hardesty, B. (1979) J. Biol. Chem., 254, 3145–3147.
91. Hunter, T., Hunt, T., Jackson, R.J. and Robertson, H.D. (1975) J. Biol. Chem., 250, 409–417.
92. Kimchi, A., Zilberstein, A., Schmidt, A., Schulman, L. and Revel, M. (1979) J. Biol. Chem., 254, 9846–9853.
93. Clemens, M.J. and Williams, B.K.G. (1978) Cell, 13, 565–572.
94. Farrell, P.J., Sen G.C., Dubois, M.F., Ratner, L., Slattery, E. and Lengyel, P. (1978) Proc. Natl. Acad. Sci. USA, 75, 5893–5898.
95. Williams, B.R.G., Gilbert, C.S. and Kerr, I.M. (1979) Nucl. Acids Res. 6, 1335–1350.
96. Jackson, R.J. (1975) MTP International Reviews of Science: Biochemistry Series One, Vol. 7, 89–136.
97. Cimadevilla, J.M., Kramer, G., Pinphanichakaru, P., Konechi, D. and Hardesty, B. (1975) Arch. Biochem. Biophys., 171, 145–153.
98. Clemens, M.J., Pain, V.M., Henshaw, E.C. and London, I.M. (1976) Biochem. Biophys. Res. Commun., 72, 768–775.
99. Delaunay, J., Ranu, R.S., Levin, D.H., Ernst, V. and London, I.M. (1977) Proc. Natl. Acad. Sci. USA, 74, 2264–2268.
100. Weber, L.A., Feman, E.R. and Baglioni, C. (1975) Biochemistry, 14, 5315–5321.
101. Thomas, G., Siegmann, M. and Gordon, J. (1979) Proc. Natl. Acad. Sci. USA, 76, 3952–3956.
102. Smith, C.J., Wejksnora, P.J., Warner, J.R., Reuben, C.S. and Rosen, O.M. (1979) Proc. Natl. Acad. Sci. USA, 76, 2725–2729.

CHAPTER 9

The control of phosphorylation of ribosomal proteins

DAVID P. LEADER

1. Introduction

The first reports of the presence of ribosomal phosphoproteins in intact cells appeared in 1970 when Kabat [1] described several phosphoproteins in ribosomes from rabbit reticulocytes, and Loeb and Blat [2] described a single phosphoprotein in ribosomes from rat liver. These reports were followed shortly afterwards by demonstrations that ribosomal proteins could be phosphorylated in vitro by either endogenous [3,4] or exogenous [5,6] protein kinases. It was widely expected that a function would soon be discovered for the phosphorylation, most probably in the regulation of ribosomal activity. However, it should be stated at the outset that this expectation has not so far been realised. Despite this, there is good reason to think that the phosphorylation of ribosomal proteins is of importance. Thus, as shall be seen, the phosphorylation is normally confined to two specific proteins of the seventy or eighty on the ribosome, these proteins are probably of functional importance, and their phosphorylations have been conserved throughout the eukaryotic kingdom.

This chapter will attempt to summarise what is known of the phosphorylation of ribosomal proteins, particularly in relation to work from the author's own laboratory. An attempt has been made to cover all the relevant literature up to the end of 1978, with reference to some later papers also included. Earlier reviews of this area can be found elsewhere [7–13].

1.1. Occurrence and number of ribosomal phosphoproteins

Ribosomal phosphoproteins are widely distributed throughout the eukaryotes. They have been detected in plants [14], and in the primitive

Cohen (ed.) Recently discovered systems of enzyme regulation by reversible phosphorylation
© *Elsevier/North-Holland Biomedical Press, 1980*

eukaryotes, *Saccharomyces cerevisiae* [15–18], *Artemia salina* [19,20], *Physarum polycephalum* [21], and *Tetrahymena pyroformis* [22,23], as well as in mammals, such as human [24,25], rabbit [1,26], rat [2,27], mouse [28,29], hamster [30,31] and bovine [32]. They are present in liver [2], kidney [33], reticulocytes [1], brain [34], adrenals [35], mammary gland [36], pituitary [32], and also in various tumour cells [28,29,37,38].

Bacterial cells are generally thought not to contain protein kinases and the bulk of evidence supports the contention that the ribosomes of *Escherichia coli* do not normally contain phosphoproteins [39–41]. There is one report of ribosomal phosphoproteins [42] and one of ribosomal protein kinase [43] in *E.coli*, but there are reasons to doubt the validity of these [44]. Several ribosomal proteins of *E.coli* do become phosphorylated to a small extent following infection of the bacterium with phage T7 (which codes for its own protein kinase), but it is not thought this phosphorylation is of functional significance [40,44]. However, this observation does demonstrate that the ribosomal proteins of *E.coli* are not dephosphorylated because of being unsuitable substrates for protein kinases and this is further shown by the finding that certain ribosomal proteins of *E.coli* can be phosphorylated by eukaryotic protein kinases in vitro [45,46,41].

Early reports of ribosomal phosphoproteins showed considerable discrepancies in the number of such proteins detected [47]. Thus different workers described eight [36], five [48,34], four [47], two [49] or only one [50,37] phosphoprotein. There are a variety of reasons for these discrepancies. Overestimates can be produced if the ribosomes are contaminated with non-ribosomal phosphoproteins [47] which co-migrate with ribosomal proteins on one-dimensional gel electrophoresis. This can occur even when the proteins have satisfied the criterion [10] of being resistant to removal from the ribosome by buffers of high ionic strength, and the further dissociation of ribosomes into their subunits at high ionic strength and temperature may be required to remove non-ribosomal contaminants [51]. The introduction of two-dimensional gel electrophoresis [52] has enabled the ribosomal proteins to be completely resolved from each other and, in many cases, to be differentiated from non-ribosomal contaminants by intensity of staining. Analysis by this method is now generally required before a phosphoprotein can be accepted as being ribosomal. However, although this is a reliable method for identifying basic ribosomal proteins, it can be difficult to apply to acidic ribosomal proteins and sometimes fails to detect them (see Section 3.). Acidic ribosomal proteins can also be lost from the ribosome under certain conditions of high ionic strength [19, 53]. It is clearly not easy, therefore, to define universally applicable criteria for establishing that a phosphoprotein is ribosomal. A clear distinction should, moreover, be made between major and minor phosphorylation of ribosomal proteins. This should preferably be done on the basis of the stoichiometry of phosphorylation, but, failing this, the relative radioactive labelling of the different proteins should be considered.

1.2. General characteristics of the ribosomal phosphoproteins of BHK and ascites cells

When ribosomal proteins are labelled with [^{32}P]orthophosphate in vivo and then subjected to one dimensional polyacrylamide gel electrophoresis in the presence of sodium dodecyl sulphate, two major phosphorylated bands are observed (Figure 1). One has a molecular weight of 31 000 and is located in the 40 S subunit, while one of a mol. wt. of approx. 14 000 is located on the 60 S subunit. The latter often appears as a closely migrating doublet and may consist of two proteins (Section 3.1). The results illustrated in Figure 1 are for baby hamster kidney fibroblasts cells (BHK), but similar results have been obtained in a variety of cell types although the relative intensities of the bands may vary (the phosphorylation of the 40 S subunit can be much greater).

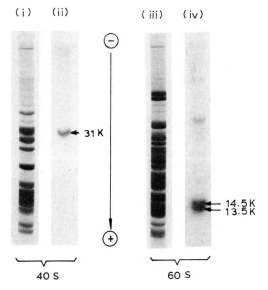

Figure 1. Sodium dodecyl sulphate gel electrophoresis of protein extracted from the ribosomal subunits of BHK cells incubated with [^{32}P]orthophosphate for 3 h, as described in [30] and [54]. Ribosomal protein (70 μg) was subjected to electrophoresis: (i) stained gel of protein from the 40 S subunit; (ii) autoradiograph of protein from the 40 S subunit; (iii) stained gel of protein from the 60 S subunit; (iv) autoradiograph of protein from the 60 S subunit.

The stoichiometry of phosphorylation calculated from the specific radio-activity of the phosphoproteins and that of the cellular ATP[47], indicates that approximately 1.4 mol of phosphate are incorporated per mole of 40 S subunit and 2.7 mol of phosphate per mole of 60 S subunit from BHK cells. This is somewhat higher than the values of 0.21 and 0.36 mol respectively, reported previously for ascites cells [47]. These lower values are comparable with those reported for *Lemna minor* [14] (0.6–0.75) and *Saccharomyces carlsbergensis* [55] (1.56), using similar methods; but are less than the values found in rat liver [2,56] (approx. 30) and rabbit reticulocytes [57] (approx. 14) measured by direct chemical analysis. Although very high values (7–11) were also found in rabbit reticulocytes by an independent method based on the rate of phosphate turnover [48], the lower values seem to accord better with the number of phosphoproteins, even allowing for multiple phosphorylations (see later).

The two major phosphoproteins of BHK cells both contain phosphoserine, with no detectable phosphothreonine. In our initial studies on ascites cells we also detected phosphothreonine [29,47], but this apparently arose from other phosphoproteins (minor species or contaminants) as it was subsequently found to be absent from both the 40 S [58] and 60 S [53] subunit of purer preparations of these ribosomes. Others had also detected phosphothreonine in certain preparations in ribosomal proteins [1,32,59,31].

2. Ribosomal protein S6

The major phosphoprotein of the 40 S subunit was the first to be identified as one of the ribosomal proteins separated by two-dimensional gel electrophoresis. Gressner and Wool in rat liver [27], followed, independently, by Stahl and Bielka in hepatoma [37], and Rankine and Leader in ascites cells [29], identified the protein as S6. (The nomenclature to be used throughout, unless stated to the contrary, is that for basic ribosomal proteins recently agreed by several different groups [60], and is an extension of the original system of Sherton and Wool [61] for proteins separated on Kaltschmidt-Wittmann gels [52]. Ribosomal proteins separated on other two-dimensional gel systems can often be correlated with those separated on Kaltschmidt-Wittmann gels by reference to a recent paper by Madjar et al. [62].) It can be seen from Figure 2 that the extent of phosphorylation of S6 can vary, and that when S6 is highly phosphorylated an anodic 'tail' of phosphorylated derivatives can be seen in the staining pattern of the gel. Under certain conditions, Gressner and Wool [27] were able to resolve this 'tail' into up to five discrete derivatives, and

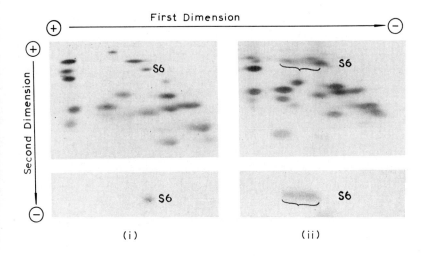

Figure 2. Two-dimensional gel electrophoresis of protein extracted from the 40 S ribosomal subunit of BHK cells incubated with [32P]orthophosphate as in Figure 1. Ribosomal protein (200 μg) was subjected to electrophoresis, as described in [52] and [63]. The upper frame of each part shows a stained gel and the lower frame the relevant area of the corresponding autoradiograph of protein from: (i) confluent cells; (ii) pre-confluent cells. See Section 2.3 for further explanation.

others have detected a similar number of components, also in rat liver [64]. As the two-dimensional gel-electrophoretic patterns of ribosomal proteins from different mammalian species are quite similar, it has been possible to conclude that, apparently without exception, the same ribosomal protein, S6, is the major phosphorylated 40 S protein in several different mammalian species and tissues [24,30,65–70]. In more primitive eukaryotes there is also agreement that there is a single, relatively large, basic phosphoprotein in the small ribosomal subunit [17,18,22,55,21]; and, although the difference in polyacrylamide gel patterns has prevented the unambiguous identification of this as S6, there is every reason to believe that the proteins are analagous.

2.1. Cyclic AMP and the phosphorylation of S6

Although ribosomal protein S6 can exist in a multiply-phosphorylated state, in most mammalian tissues it is normally phosphorylated to a much lower extent. One type of stimulus which has been found to promote extensive phosphorylation of S6 is cyclic AMP or agents which cause increases in the intracellular concentrations of cyclic AMP. This is especially true of rat liver, in which glucagon was shown to increase the 32P-radioactivity associated with a 40 S ribosomal protein [71]. Gressner and Wool subsequently showed that glucagon caused the multiple phosphorylation of ribosomal protein S6 [72], and observed similar effects when rats were injected with cyclic AMP or its dibutyryl derivative [72], or when rats were made diabetic by injection of streptozotocin [73]. A similar response was observed following anaesthetic or operative stress [64], and this was probably due to an increased secretion of adrenalin. Phosphorylation of ribosomal protein S6 in rat cerebral cortex [67] is stimulated by cyclic AMP (but not by cyclic GMP); and the phosphorylation of proteins which most likely correspond to S6 has also been demonstrated in bovine pituitary [32], rabbit reticulocytes [59] and rat pancreatic islet tumour cells [74]. ACTH and cyclic AMP increase the incorporation of ^{32}P-radioactivity into mouse adrenal ribosomes [35], while the phosphate content of rat liver ribosomes is decreased by thyroidectomy [56] (a treatment which may decrease the concentration of cyclic AMP). It should however be mentioned that in none of these studies is it clear whether increased phosphorylation of S6 in response to cyclic AMP is caused by an activation of cyclic AMP-dependent protein kinase, an inhibition of S6 phosphatase or both.

In contrast to those mammalian tissues which respond to cyclic AMP, in other cells the phosphorylation of S6 does not appear to increase in response to this cyclic nucleotide. This is true of ascites cells [75], glioma cells [76] and yeast [17]. Moreover, although an effect of cyclic AMP has been observed in BHK cells [77], like that in reticulocytes [59,1] and HeLa cells [78], it is difficult to obtain reproducibly [202].

2.2. Effect of inhibitors of protein synthesis on the state of phosphorylation of S6

Some antimetabolites known to inhibit protein biosynthesis stimulate the phosphorylation of S6. Kabat found that sodium fluoride increased the content of [^{32}P]orthophosphate of the major phosphoprotein (F or II) in the 40 S ribosomal subunit of rabbit reticulocytes (presumably S6) [1], and the appearance of multiply phosphorylated derivatives of S6 was subsequently observed by Gressner and Wool in the livers of rats that had been treated with puromycin or cycloheximide [27]. The effect of cycloheximide has also been observed in HeLa cells [79]. Fusidic acid has been reported to stimulate the incorporation of [^{32}P]orthophosphate into the total ribosomal protein of rat liver [80], and a similar effect of aurintricarboxylic acid on a 40 S ribosomal protein of L-cells has been reported [81]. However, the phosphorylation of ribosomal protein S6 was not stimulated by cycloheximide in yeast [17], nor by sodium fluoride in *Tetrahymena* [22], nor by cycloheximide, pactamycin or sodium fluoride in ascites cells [82,83]. Since many of these inhibitors are known to influence the metabolism of cyclic AMP [84–87], the effect of these compounds may be quite unrelated to their effects on protein synthesis, and the variable response by different cells may reflect differences in their response to cyclic AMP. Other possibilities are not however excluded; for example the synthesis of a rapidly turning over phosphoprotein phosphatase might be inhibited. In addition, sodium fluoride is well known to be a powerful inhibitor of many protein phosphatases.

The phosphorylation of S6 is also stimulated by a number of agents which cause liver injury and inhibit protein biosynthesis. However, in these cases there appear to be no alterations in cyclic nucleotide metabolism. These agents include D-galactosamine [88], thioacetamide [89], and dimethylnitrosamine [90]. Increased phosphorylation of S6 has been observed when Tetrahymena cells are deprived of nutrients [22], and a similar effect has been claimed in yeast cells [91]. However, these latter results are quite different from those obtained when mammalian cells are deprived of nutrients: in these cells dephosphorylation is more often observed [30,25,70].

There have been some reports that the phosphorylation of S6 is increased when animal cells are infected by certain viruses which inhibit the synthesis of host cell proteins. It is certainly true that the incorporation of [^{32}P]orthophosphate into S6 is increased following infection of HeLa cells with vaccinia virus [24,79] or adenovirus [92,93], and of Ehrlich ascites cells with mengovirus [68]; although not following infection of L-cells with vesicular stomatitis virus [66]. However, as neither the number nor the relative proportion of phosphorylated derivatives of S6 appear to be grossly affected, it is possible that the viruses merely stimulate the uptake of [^{32}P]orthophosphate by the cells [94].

2.3. Increased phosphorylation of S6 during stimulation of cellular proliferation

Although the phosphorylation of S6 in cultivated mammalian cells is often difficult to promote with exogenous cyclic AMP, it can be markedly enhanced during rapid cellular proliferation. Thus rapidly growing, preconfluent, BHK cell monolayers contain ribosomal protein S6 in the multiply-phosphorylated state, in contrast to cells which have reached confluence (Fig.2) [30]. This study was also the first demonstration that the phosphorylation of S6 could be extensive without there being elevated cellular concentrations of cyclic AMP. However, since the cells were asynchronous, fluctuations in the concentration of cyclic AMP, which are known to occur during the cell cycle [95], might have been obscured. A role for cyclic AMP is therefore not excluded, although it is worth pointing out that the conditions under which the phosphorylation was observed are the opposite to those which are associated with raised intracellular concentrations of cyclic AMP in tissue culture cells [96]. Another situation in which the phosphorylation of S6 increases during cellular proliferation, occurs when HeLa cells are resuspended in fresh medium [25], and in this case insulin and amino acids are required for the stimulation of phosphorylation [78]. Increased phosphorylation of S6 also takes place when serum is restored to stationary chick embryo fibroblasts, causing them to progress into the G_2 phase of the cell cycle [70]. It is interesting to note in this context, that in one of the earliest studies of ribosome phosphorylation, it was reported that cells grown on medium containing serum had a higher ribosomal protein kinase activity than cells maintained on medium lacking serum [4].

There is a discrepancy between our own results and the others mentioned above. We were unable to detect an increase in the phosphorylation of S6 when confluent BHK cells were stimulated to grow again by renewal of their medium [30]. A possible reason for this is the high cell density in our experiments: Lastick et al. found that the phosphorylation of ribosomal protein S6 in HeLa cells was not stimulated if these were resuspended in fresh medium at too high a cell density [25]. Why the cell density should be so crucial is unclear.

Several workers have found extensive phosphorylation of S6 in regenerating rat liver [50,97–99]. There are, however, increases in the concentration of cyclic AMP during liver regeneration and the stimulus for this phosphorylation is therefore unclear [100]. Another effect which could be related to those described above is the decreased phosphorylation of a 40 S ribosomal protein in Lemna minor caused by addition of the plant-growth-inhibiting hormone, abscisic acid [14].

Although the stimulated cells in the studies cited above had higher rates of protein synthesis than the unstimulated cells, the correlation between phosphorylation of S6 and protein synthesis was not stringent [30,25]. Nevertheless it is striking that several different workers have found that S6 is much more highly phosphorylated on polysomes than on monosomes. This has been found in rabbit reticulocytes [1,48], sarcoma cells [28], BHK cells [77] (see Figure 3) and myeloma cells [69]. In some other studies such differences were either small [14], or absent [50]; however, because of the possibility of cross-contamination between the polysome and monosome fractions (especially in tissues such as rat liver, where polysomes predominate)

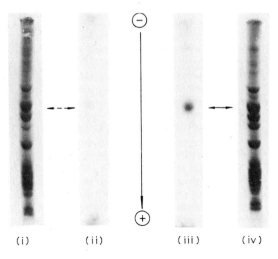

(i)   (ii)   (iii)   (iv)

Figure 3. Sodium dodecyl sulphate gel electrophoresis of protein extracted from the 40 S ribosomal subunits of BHK cells incubated with [^{32}P]orthophosphate for 3 h and fractionated into monoribosomes and polyribosomes, as described in [77]. Ribosomal protein (70 μg) was subjected to electrophoresis: (i) stained gel of protein from monoribosomes; (ii) autoradiograph of protein from monoribosomes; (iii) autoradiograph of protein from polyribosomes; (iv) stained gel of protein from polyribosomes.

one is inclined to attach more weight to the studies where differences were apparent. Moreover the greater phosphorylation of membrane-bound ribosomes compared with free ribosomes may be due to their containing a greater proportion of polysomes [49,101,102]. Furthermore, the dephosphorylation of the 38 000 molecular weight protein (probably S6) of CHO cells at mitosis might have been related to a decreased number of polysomes [31].

2.4. Structure of S6 and possible functions for its phosphorylation

Wool and coworkers have purified S6 and found it to have a molecular weight of about 31 000 [103]. They have also started to determine its primary structure [11] in order to identify the sites of phosphorylation. The ability of S6 to serve as a substrate for protein kinase(s) in vivo and in vitro [37,104,65] clearly indicates that it is at least partially on the surface of the 40 S subunit, although some studies with chemical probes suggest a large proportion of the protein is buried beneath the surface [105,106]. On the other hand these latter results are in conflict with iodination studies using lactoperoxidase [63], and are also somewhat difficult to reconcile with the finding that, in yeast, the protein corresponding to S6 appears to be one of the last ones to be added during the assembly process [55,107]. There is some reason to believe that in the 80 S ribosome S6 is located at the interface between the subunits, for in yeast [17] and *Tetrahymena* [22] the phosphorylated protein transfers to the 60 S subunit under certain preparative conditions, and this is consistent with a masking of the protein to enzymic probes following the formation of the 80 S ribosome [37,108,99]. The greater susceptibility to dephosphorylation in the 40 S rather than the 80 S ribosome is also consistent with this idea [109].

Ribosomal protein S6 has been found to bind to tRNA-sepharose [110], suggesting that it might interact with tRNA; possibly the initiator tRNA in view of the report that S6 can be crosslinked to the initiation factor eIF-2 (see Chapter 8) and that an antiserum to S6 inhibits the binding of the ternary initiation complex to the 40 S ribosomal subunit [111]. S6 may also interact with mRNA [112], and (despite earlier indications to the contrary [10]) it appears to cross-react with an antiserum raised against the ribosomal proteins of *E.coli* [113] implying extraordinarily strong evolutionary constraints on its structure.

These latter reports must surely give considerable direction to studies on the function of the phosphorylation of S6. However, before discussing possible functions for the protein it is necessary to emphasize that the nature of the protein kinase(s) which phosphorylates S6 in vivo is not yet clear. S6 is phosphorylated by cyclic AMP-dependent protein kinase in vitro, but it should be evident from Sections 2.1–2.3 that other protein kinases (controlled in some other manner) may well be involved. The possibility cannot be excluded that multiple protein kinases are responsible for the phosphorylation of the different phosphorylation sites on S6, and several examples of this phenomenon are now known (see Chapters 1, 3, 8 and 11 of this book). This point is of importance as it raises the possibility (unpleasant in its complexity) that the phosphorylation of S6 in different circumstances could, by involving different sites, have different functional consequences.

Although the idea that the phosphorylation of S6 plays a role in the regulation of protein biosynthesis is aesthetically pleasing, it has been difficult to reconcile with the experimental evidence. Thus, S6 has been found to be phosphorylated in response to agents which both inhibit (Section 2.2) and stimulate (Section 2.3) protein synthesis. Furthermore, those situations in which the phosphorylation of S6 is stimulated by cyclic AMP do not fit readily into such a scheme. Cyclic AMP has been reported to stimulate [114–116], to have no affect on [117,76], or to inhibit [118–122] protein synthesis in vitro in various tissues. However, in the liver (the main tissue in which cyclic AMP increases the phosphorylation of S6), there is little or no effect of cyclic AMP on total protein synthesis, although induction of specific enzymes may occur [123–126]. This is also true of short-term experimental diabetes, which has little affect on the total liver protein synthesis [127,128], although it severely inhibits skeletal muscle protein synthesis [129].

Other possible functions for the phosphorylation of S6 have been considered but these, in turn, attract little support. Thus the idea [30] that the phosphorylation of S6 might be important for nuclear assembly and extranuclear transport of ribosomes is not consistent with either our own work, (Kennedy and Leader, submitted [203]) or other studies of the phosphorylation state of newly synthesized cytoplasmic [25] or nucleolar [130] S6. It has been suggested that the phosphorylation of S6 protects ribosomes from degradation. Although this idea is consistent with the lower phosphorylation of S6 in resting cells (Section 2.3), where ribosome degradation increases [131], it is difficult to relate to the effects of cyclic AMP on S6 phosphorylation, and is directly opposed by evidence from *Tetrahymena* [22,23,132]. However as mentioned earlier, S6 may be phosphorylated by protein kinases distinct from cyclic AMP-dependent protein kinase, and in view of the evidence for translational control in cells that have been stimulated to proliferate [133], the possibility that S6 phosphorylation modulates the rate of protein synthesis needs to be reconsidered.

Clearly the fact that ribosomes such as those of normal rat liver are minimally phosphorylated whilst being very active in protein biosynthesis means that phosphorylation of S6 is not essential for ribosomes to be active in protein biosynthesis. Nevertheless, it is possible that the phosphorylation of S6 might have a conditional influence on protein biosynthesis. The phosphorylation of many enzymes alters the K_m for substrates and effectors without affecting the V_{max} (see Chapters 1–3). Therefore by analogy, the phosphorylation of S6 might increase the affinity of the ribosome for another molecule(s) involved in protein synthesis, and the overall rate of protein synthesis might only be influenced if the concentration of the molecule in question were sufficiently low. One could then rationalise phosphorylation of S6 as both a cellular response to rapid proliferation, where the protein synthetic capacity of the cell is suddenly put under pressure; and as a compensation mechanism brought into play following cellular damage, when there is again a need to optimise the use of diminished resources. Moreover one could argue that the situations where cyclic AMP enhances the phosphorylation of S6 in the liver, and perhaps other tissues, without a marked effect on protein synthesis, are situations where the rate of protein synthesis is not being limited by the concentration of the putative key components.

214

In order to investigate the role of S6 phosphorylation in protein synthesis, experiments must meet the following criteria.

1. The ribosomes must be compared under conditions where the state of phosphorylation of S6 (and S6 alone) is significantly different;
2. The factors and enzymes used for protein synthesis must not contain protein kinases or protein phosphatases;
3. The initiation reactions of protein synthesis on a natural mRNA should be studied, as well as the elongation reactions;
4. The experiments should examine the rate of translation at limiting concentrations of the different components.

Unfortunately all these criteria have never been met. Thus the two studies which reported an inhibition of protein synthesis [120,122] contained no evidence that treatment of ribosomes with protein kinase changed the extent of phosphorylation of ribosomal protein S6 and it alone. Several of the studies which failed to detect a difference in the activity of the ribosomes [37,98,15,16] merely measured polyphenylalanine synthesis, and used preparations of elongation factors which were not known to be free of phosphoprotein kinases and phosphatases. The studies of Eil and Wool [134] and Kabat and co-workers [7,8], which also failed to detect changes in the rate of protein synthesis, were far more thorough, and employed natural mRNAs. However, Kabat and coworkers did not use factors purified free of kinases and phosphatases, and were not able to vary the concentrations of individual factors independently. These criticisms cannot so easily be made against the work of Eil and Wool, as these workers used partially resolved factors, purified free of phosphoprotein phosphatase. However, the major deficiency of their work was the use of exogenous cyclic AMP-dependent protein kinase to produce phosphorylated ribosomes. This phosphorylates other ribosomal proteins besides S6 in vitro [135,104]; and in addition the final extent of phosphorylation of S6 in these experiments was unknown. Moreover this enzyme may not be the relevant protein kinase to test. It would therefore seem that a careful re-examination of the effect of phosphorylation of S6 on the protein synthetic activity of ribosomes, fulfilling the above criteria, is required.

3. The major phosphoprotein of the 60 S ribosomal subunit

In the original studies of Kabat and coworkers, a phosphoprotein (Si or I) of relatively low molecular weight was detected by one-dimensional polyacrylamide gel electrophoresis of the ribosomal proteins of the 60 S subunit of reticulocytes or sarcoma cells [59]. The labelling of this protein by

[^{32}P]orthophosphate was similar to the basal level of labelling of the phosphoprotein on the 40 S subunit (F or II), now known to have been ribosomal protein S6. We subsequently found a similar phosphoprotein in the ribosomes of ascites cells [83,136,47] and BHK cells (Figure 1). Like Kabat's band I, this was of relatively low molecular weight, normally found on the 60 S ribosomal subunit, and could sometimes be resolved into two components. The extent of labelling of this protein in ribosomal subunits prepared under conditions which remove non-ribosomal contaminants, together with a large body of evidence demonstrating its phosphoprotein nature, indicated that there was another phosphoprotein on the ribosome, besides S6.

Nevertheless there was, for a long time, reluctance to recognise the existence of a second ribosomal phosphoprotein, mainly because of the difficulty in radioactively labelling the protein in rat liver [50], where incubation with alkaline phosphatase and electrophoretic analysis was eventually required to establish its phosphorylation [137]. Our own endeavours to detect the 60 S phosphoprotein on two-dimensional gels were influenced by indications from acid urea gels that the protein might be more acidic than the bulk of ribosomal proteins [47]. Indeed, Martini and Gould [138] had observed the phosphorylation of a relatively acidic, low molecular weight, 60 S ribosomal protein, by an endogenous protein kinase following incubation with [γ-^{32}P]ATP. Initially we found an apparently neutral radioactive phosphoprotein on our two-dimensional gels, which we designated Lγ [47] (a similar observation was made in yeast [18]). Later an extremely acidic, weakly staining, 60 S phosphoprotein of similar molecular weight to Lγ was detected [139]. This suggested that our apparently neutral protein, Lγ, might in reality be an aggregated form of an acidic protein [139] and this possibility was strengthened by the finding of Zinker and Warner that heavily labelled acidic ribosomal phosphoproteins were present in yeast [17].

3.1. Evidence for an acidic ribosomal phosphoprotein

Other workers, besides ourselves, have found difficulty in detecting acidic ribosomal proteins using the standard Kaltschmidt-Wittmann method of two-dimensional gel electrophoresis [140–142]. We therefore devised a modified method of gel electrophoresis to allow better detection and analysis of the acidic phosphoprotein, and this method ('sweep' gel electrophoresis) is described in detail elsewhere [54]. The essential feature is the establishment of a difference in pH between the upper electrophoresis buffer (pH 3.55) and the first dimensional gel (pH 5.5). This causes acidic proteins of isoelectric points between pH 3.55 and 5.5 to be swept into the gel in a concentrated band, some distance behind the more rapidly migrating basic proteins. The gels are then subjected to electrophoresis in a second dimension (18% acrylamide, pH 4.5) where proteins are predominantly separated on the basis of size (although charge may also be a factor in the case of acidic proteins [143]). A representative gel from ascites cells labelled with [^{32}P]orthophosphate is shown in Figure 4, and it can be seen that at least two stained spots, one of which is highly labelled, are seen on the 'sweep' front, but *no* neutral phosphoprotein is present. The position of the spots can be compared in different experiments by reference to the basic proteins, and after analysing many gels we have concluded that there are a maximum of three spots which stain for protein, only the upper two of which are ever labelled [54,143,53].

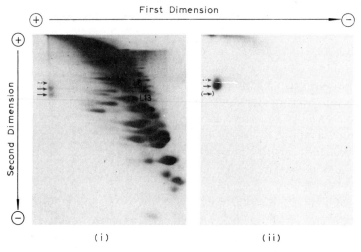

Figure 4. Two-dimensional 'sweep' gel electrophoresis of protein extracted from the 60 S ribosomal subunit of ascites cells incubated with [^{32}P]orthophosphate, as described in [29]. Ribosomal protein (200 μg) was subjected to electrophoresis, as described in [54]: (i) stained gel; (ii) autoradiograph. The arrows indicate corresponding positions on the two frames, a broken arrow denoting a weak spot and an arrow in parentheses indicating the absence of a spot.

Our conclusion that the major 60 S phosphoprotein of ascites cells (and BHK cells) is an acidic protein is supported by isoelectric focusing experiments in gels, which suggest an isoelectric point of pH 4.4–4.5 [53]. We were unable to incorporate sufficient radioactivity into rat liver ribosomal proteins to test whether the analogous protein in this tissue was phosphorylated. However, as a single protein staining spot corresponding to the uppermost position in Figure 4 predominated, we concluded [53] that in the liver Lγ was present as the more slowly migrating phosphoprotein species. This seems to be the case in many cells growing under good conditions [54,143,53].

Following the initial reports of phosphorylated acidic ribosomal proteins by Zinker and Warner [17] and ourselves [139], ample confirmation has been provided by others in studies with *A. salina* [19,20], HeLa cells [144], rat liver [145,137], yeast [91,107], brain [67] and L-cells [146]. There are some differences between the number of stained and radioactive spots observed by different workers, but most workers observe a maximum of either two [17,19,91,107,146] or three [54,145,144] closely migrating protein staining spots, not more than two of which are labelled. Some have concluded that there is a single phosphorylation site per molecule [54,19], whereas others favour the existence of two such sites [144,145]. Purification of the acidic proteins has provided more information on this point. It seems almost certain that there is only a single phosphoserine residue in the major acidic ribosomal phosphoprotein (eL12) of *A. salina* [20], although it appears that more phosphoryl residues are present in rat liver. Möller and co-workers found up to two residues in their 'eL12' [20], Wool and coworkers up to three residues in their 'P2' [137], and Vásquez and co-workers, using a partially purified preparation, suggested up to three or four residues [147].

In addition to the acidic phosphoprotein of the 60 S ribosomal subunit, discussed above, there appears to be a second acidic phosphoprotein that has generally (although perhaps incorrectly) been thought of as a minor species. This was first observed as a series of more weakly staining 'satellites' of the 'major' phosphorylated protein and its derivatives. These satellites, which are considerably more acidic and slightly larger than eL12 [17,19,137], have been purified by Wool [137] and Möller [19,20]. Although it was initially suspected that the smaller acidic proteins might be derived from the larger ones by proteolysis, it is now clear that the two sets of proteins are distinct gene products [20]. Möller and co-workers have recently reported that the larger acidic phosphoprotein (eL12′) can be recovered in about 50% of the yield of the smaller protein (eL12) [20], and this suggests that it may not be such a minor species as originally indicated by two-dimensional gel analysis of total ribosomal protein. It is therefore possible that this protein is poorly soluble in the buffers used for the two-dimensional gels, and that the 14.5 K and 13.5 K phosphoproteins of the 60 S ribosome observed by one-dimensional SDS gel electrophoresis (Figure 1) corresponded to eL12′ and eL12. In this case the 'sweep' gels (Figure 4) may only be visualizing eL12, and the two ^{32}P-labelled spots that are observed correspond to two phosphorylated derivatives of this protein.

218

3.2. Relationship to proteins L7/L12 of E.coli

Much of the interest in the acidic ribosomal phosphoproteins of eukaryotes relates to the possibility that these are the eukaryotic equivalent of the acidic ribosomal proteins, L7 and L12, of *E.coli*. The proteins L7 and L12 differ only in the amino terminal *N*-acetyl group on L7 [148], are thought to be present in four copies per ribosome [149] (all other ribosomal proteins occur as single copies), and have been strongly implicated in the GTPase reactions of the ribosome [150]. It became apparent that these prokaryotic acidic ribosomal proteins have counterparts in eukaryotes when anti-serum raised against them cross-reacted with 60 S ribosomal subunits from yeast [10], rat liver [10,151] and chicken liver [152]; and inhibited eukaryotic poly-phenylalanine synthesis [152,153]. Furthermore, the ribosomal proteins L7 and L12 of *E.coli* are selectively removed from the ribosome with ethanol and ammonium chloride [154], and when similar treatment was applied to ribosomes from yeast [10,155], rat liver [151] or *A.salina* [156], acidic proteins were extracted which could functionally replace (in the case of yeast [10]) or be replaced by (in the cases of yeast [155] and *A.salina* [156]) *E.coli* proteins L7/L12.

The question of most relevance to this chapter is whether these acidic proteins correspond to the phosphorylated acidic proteins discussed in Section 3.1. Unfortunately, though much of the evidence suggests that this is the case, puzzling discrepancies prevent a definite answer. Thus the proteins eL12 and eL12′ from *A.salina*, referred to above, have been purified to homogeneity and are, indeed, phosphorylated [20]. However, despite the functional exchangeability with *E.coli* L7/L12 observed previously [156], eL12 and eL12′ did not cross-react immunochemically with *E.coli* proteins L7/L12 [19], and the amino acid sequence homology between the prokaryotic and eukaryotic proteins is very low [157,158]. Stöffler and Wool, on the other hand, were able to isolate proteins from rat liver 60 S ribosomal subunits that did cross-react immunochemically with proteins L7/L12 of *E.coli* [151]. These were designated L40 and L41 and were detected in other tissues [159]. However, this work was done before the question of the phosphorylation of acidic ribosomal proteins arose, and more recent attempts by Wool and coworkers to isolate these proteins have been unsuccessful. In the process of purifying almost all of the proteins of rat liver ribosomes, Wool and co-workers did isolate acidic phosphoproteins (P1 and P2 and derivatives) [137]. Although they believe that these are not equivalent to L40 and L41 as they do not cross-react with L7/L12 of *E.coli*, doubt will persist until L40 and L41 are rediscovered. The acidic proteins studied by Wool's group and Möller's group do appear to be analogous. Thus rat liver protein P2 appears to correspond to eL12 of *A.salina* on the basis of similarity in N-terminal amino acid sequence, even though the two proteins were reported not to cross-react immunochemically [11,157].

Our own results lend support to a relationship between the eukaryotic acidic phosphoprotein (Lγ) and the proteins L7/L12 of *E.coli* [53]. We have raised antibodies against pieces of two-dimensional gel containing Lγ and found these to cross-react with *E.coli* proteins L7/L12 in Ochterlony immunodiffusion. These immunochemical results are not conclusive, however, for our preparations of Lγ were not sufficiently pure to enable us to perform the reciprocal experiment, and we have been unable, as yet, to obtain cross-reaction between Lγ from ascites cells and antiserum raised against *E.coli* L7/L12.

In conclusion it would appear that earlier immunochemical and functional interchangeability studies between prokaryotic and eukaryotic acidic ribosomal proteins are not as conclusive as they first appeared. There have been no recent reports confirming the interchangeability studies, and one gathers that this is not from want of trying. The immunochemical results seem almost to indicate that success is only obtained with unfractionated ribosomes or proteins, or proteins presented in polyacrylamide; but whether this can be explained in terms of protein structure or immunochemical artefact is unclear. It should, however, be admitted that immunochemical cross-reactivity would not be expected between proteins with as little similarity in primary structure as L7/L12 of *E.coli* [148] and eL12 of *A.salina* [158]. One is thus obliged to fall back largely on the structural similarities between the acidic proteins from prokaryotes and eukaryotes: their sizes, isoelectric points, probable location at the interface between the subunits [160,136,144] and susceptibility to phosphorylation by protein kinase in vitro [46,41]. A more compelling point of structural similarity is the fact that the acidic phosphoproteins from *A.salina* [20] and yeast [55,161], like L7/L12 of *E.coli*, are present in multiple copies. Finally there is some reason to believe that the eukaryotic acidic phosphoproteins may perform functions on the ribosome similar to those served by *E.coli* L7/L12; for anti-serum against *A.salina* eL12 has been found to inhibit the EF2-dependent GTPase activity of ribosomes from *A.salina* [19].

3.3. The function of the phosphorylation of the acidic ribosomal protein

As with ribosomal protein S6, the function of the acidic ribosomal protein is, at the moment, only a subject of speculation. If one does consider that the protein is functionally analogous to L7/L12 of *E.coli*, one possibility is that the phosphorylated form of the protein is an intermediate in the GTPase reactions of protein synthesis in which it participates. However, although some support is lent to this idea by Zinker and Warner's finding that cycloheximide prevented the phosphorylation of the acidic protein in yeast [17], we did not find such an effect in ascites cells [82]. Futhermore, the slow rate of turnover of the phosphoryl groups (about 3% per min.) would appear to exclude a cycle of phosphorylation and dephosphorylation occurring with each round of translation [48].

A different function for the phosphorylation of the acidic protein was suggested in Kabat's original paper on ribosome phosphorylation [1]. In rabbit reticulocytes this protein was only found to be labelled with [^{32}P]orthophosphate on single ribosomes (monosomes), and not (or to a much lesser extent) on polysomes. As most of the monosomes in ribosome populations are inactive in protein synthesis it appeared possible that phosphorylation of the acidic protein might cause inactivation of ribosomes. However, when Bitte and Kabat examined sarcoma cell ribosomes this difference in labelling was not seen [28]. We re-examined this question in BHK cells, using the 'sweep' gel analysis of Lγ to supplement measurement of radioactivity, and also found no difference in phosphorylation between polysomes and monosomes [54]. Similarly, in the skeletal muscle of normal and diabetic rats, where the proportion of monosomes differs markedly, we found no difference in the phosphorylation of Lγ as judged from the two-dimensional gel pattern [143]. None of the other relevant published studies appear to support the original idea of Kabat. Martini and Gould found that the phosphorylation of what was probably the same acidic protein (their L28) by endogenous protein kinase and [γ-^{32}P]ATP in vitro was similar on polysomes and monosomes [138]. Jolicoeur et al. observed a phosphoprotein of 13 000 molecular weight in the ribosomes from Landschutz tumour cells, but found that inhibition of protein biosynthesis by removal of amino acids from the growth medium decreased, rather than increased, its labelling [38].

Houston examined the ability of 60 S ribosomal subunits to catalyse polyphenylalanine synthesis after dephosphorylation of the acidic protein by alkaline phosphatase, but found that this treatment did not alter the activity of the ribosomes [146]. However, as no measures were taken to exclude protein kinases from the protein synthesis factors used, the results cannot be regarded as conclusive.

Our own observations, using 'sweep' gel electrophoresis, imply that Lγ is generally in a state of almost complete phosphorylation [54,143,53] and this suggests that the phosphorylation may not have a dynamic regulatory function.

Finally, it should be mentioned that exchange of the acidic protein between the ribosomes and the cytosol fraction of yeast has been observed [17]. It is possible that the acidic protein cycles on and off the ribosome in vivo, and that this cycling, whatever its function, is somehow influenced by the state of phosphorylation of the protein.

4. Other ribosomal phosphoproteins

S6 and Lγ are the most phosphorylated proteins of eukaryotic ribosomes in cells under normal conditions, their relative contribution to the total phosphate on the ribosome depending on the degree of phosphorylation of S6 and the extent to which Lγ is lost from the ribosomes during isolation. Some other proteins are labelled very weakly with [^{32}P]orthophosphate in vivo, the amount of phosphate they contain being generally less than 0.1 mol/mol of ribosomes. However, under special circumstances, the extent of phosphorylation of certain of these proteins may increase to levels comparable with S6 and Lγ.

4.1. Ribosomal proteins S3 and L14

In our studies on the phosphorylation of ribosomes from ascites cells, we normally incubated the cells in medium lacking glucose (and amino acids) to prevent extensive glycolysis with resulting acidification of the medium. The pattern of ribosomal protein phosphorylation observed under these conditions was, with occasional exceptions (see Section 4.2 below), similar to that observed in BHK cells, and in ascites cells labelled in the peritonea of the mice, S6 and Lγ being the major phosphoproteins [29,47,54]. When full Eagle's medium containing glucose and amino acids was used for incubating the cells, ribosomal proteins did become phosphorylated, despite the resulting acidity. However, there was a completely different set of proteins phosphorylated in this case. Such a changed pattern of phosphorylation was also observed when the cells were incubated in medium lacking glucose but supplemented with amino acids, and the proteins phosphorylated were shown to be S3 (Figure 5) and L14 (Figure 6) [58]. Ribosomal protein S6 was no longer phosphorylated (Figure 5), and the extent of phosphorylation of Lγ was shown to be only a fraction of normal (Figure 7) [143]. The extent of phosphorylation of S3 and L14 appeared from the autoradiograph to be comparable with that normally found for S6 and Lγ, and this was confirmed by calculation of the stoichiometry from the specific radioactivity of ATP: in

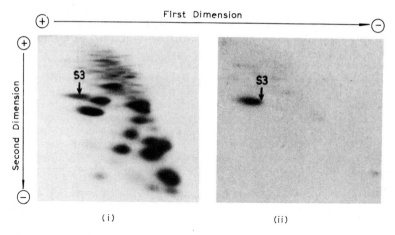

Figure 5. Two-dimensional gel electrophoresis of protein extracted from the 40 S ribosomal subunit of ascites cells incubated with [^{32}P]orthophosphate in Earle's saline, lacking glucose but supplemented with amino acids, as described in [58]. Ribosomal protein (200 μg) was subjected to electrophoresis as in Figure 2, but with a pH of 6.6 in the first dimension [47]. (i) Stained gel; (ii) autoradiograph.

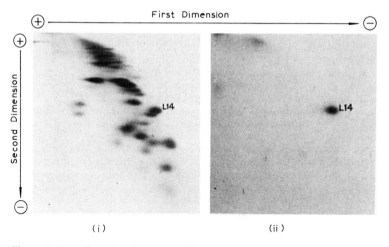

Figure 6. Two-dimensional gel electrophoresis of protein extracted from the 60 S ribosomal subunit of ascites cells incubated with [^{32}P]orthophosphate, as in Figure 5. (i) Stained gel; (ii) autoradiograph,

three experiments, an average of 0.29 mol of phosphate was incorporated into the 40 S subunit and an average of 0.56 mol of phosphate into the 60 S subunit. Only phosphoserine was detected on the 60 S subunit, but similar

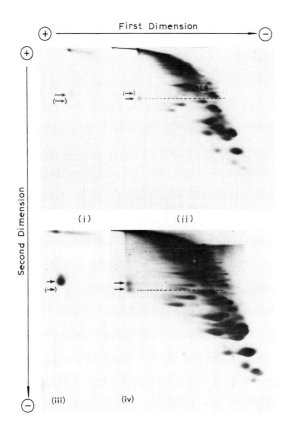

Figure 7. Two-dimensional 'sweep' gel electrophoresis of protein extracted from the 60 S ribosomal subunits of ascites cell incubated with [^{32}P]orthophosphate. The protein (200 μg) analysed had been extracted from: (i) cells incubated in Earle's saline, lacking glucose but supplemented with amino acids − section of autoradiograph corresponding to the left-hand side of (ii) stained gel; (iii) cells incubated in Earle's saline without glucose or amino acids − section of autoradiograph corresponding to left-hand side of (iv) stained gel. See legend of Figure 4 for explanation of arrows.

amounts of phosphoserine and phosphothreonine were observed on the 40 S subunit [58].

Others have also detected phosphorylation of S3 [65−67] and L14 [25,65,66,146], or proteins likely to correspond to these [18,74,55], in eukaryotic cells. However, in contrast to the results described above, there was never more than a trace of phosphate present in the proteins in these other studies. The report by Anderson et al. [97] of altered positions of migration of S3 in unlabelled regenerating rat liver is relevant here, as if this alteration were

224

caused by a change in phosphorylation state of the protein it would imply extensive phosphorylation of S3. However this altered migration is probably caused by sulphydryl oxidation [202].

The precise stimulus for the phosphorylation of S3 and L14 in ascites cells is unclear. We thought for a time that lack of glucose and persistence of protein synthesis were the common conditions producing this pattern of phosphorylation, the glucose added in our studies with full Eagle's medium being removed by glycolysis [58]. However, we have been unable to produce an alteration in the pattern of phosphorylation of the ribosomes of BHK cells by starvation for glucose, even when 2-deoxyglucose was added; and others who incubated Landschutz ascites cells without glucose, in either the presence or absence of amino acids, observed no comparable effects [38]. We now think that the inducing stimulus may be more complex: perhaps the extreme glycolytic metabolism of ascites cells is a key factor here. It may be more than mere coincidence that some phosphorylation of S3 [62] was also observed when insulinoma cells were incubated in medium containing glucose [74].

Although ribosomal protein L14 has been isolated [137], nothing is yet known of its function. A little is known about the function of S3, although care is needed in considering the data regarding S3 because the original two-dimensional gel spot S3 [61] has now been shown to contain a second protein, termed S3a [162] (See also ref. [60]). Molecular weight considerations lead us to conclude that our phosphoprotein is S3, as it is smaller than S6 [58,103], whereas S3a is larger than S6 [162]. However, direct confirmation of this point is still required. Westermann et al. [163] have cross-linked S3 (their S2) to eIF2 (the eukaryotic initiation factor that binds the initiator tRNA to the 40 S ribosomal subunit), and an anti-serum raised against S3 prevented an eIF2-containing ternary complex from binding to the 40 S subunit [164]. It is therefore tempting to speculate that the phosphorylation of S3 might interfere with this initiation reaction, but this possibility has not yet been tested.

4.2. Ribosomal proteins S2 and S16

There is another ribosomal protein that becomes extensively phosphorylated under abnormal conditions and this, in fact, was discovered before the phosphorylation of S3 and L14. Kaerlein and Horak observed that, although ribosomal protein S2 was unphosphorylated in normal HeLa cells, infection with vaccinia virus caused a phosphorylation of this protein, as great as that of S6 [24]. Another protein also became phosphorylated to about one-third the extent of the others, and this was later concluded to be S16 [79]. There appear to be two phosphorylated derivatives of S2, one containing only phosphoserine, and the other also containing phosphothreonine; and it is assumed that the phosphorylations are catalysed by a viral protein kinase [79].

This effect of vaccinia virus is not found in cells infected with adenovirus [92,93], vesicular stomatitis virus [66], mengovirus [68] or herpes virus (Kennedy and Leader, unpublished). However there is sometimes phosphorylation of ribosomal protein S2 in ascites cells incubated in a minimal medium, lacking glucose and amino acids [47]; and when the cells are incubated in full Eagle's medium (with consequent glycolysis and acidification) some phosphorylation of S2 also occurs [58]. Trace phosphorylation of ribosomal protein S2 has been detected in L-cells [66] and brain [67], but there appear to be no other reports of the phosphorylation of ribosomal protein S16. Nothing is known of the function of these ribosomal proteins and there is no evidence that their phosphorylation affects the activity of the ribosome.

4.3. Other phosphoproteins

This section will consider various claims that significant phosphorylation of other ribosomal proteins occurs. It will be necessary to restrict discussion to those cases in which the phosphoproteins have been identified by two-dimensional gel electrophoresis, despite the possible limitations of this method mentioned in Section 1.1. For this reason the reader is merely referred to two of the more interesting reports of additional phosphoproteins detected by one-dimensional gel electrophoresis [59,31].

Trace amounts of [^{32}P]orthophosphate in other ribosomal proteins have been detected on two-dimensional gels, and these proteins include S4 [66], S12 [146], L1 [146], L2 [146], L6 [58,67], L13 [67] and L22 [66]. However it can be argued that, although only weakly labelled, these proteins may, in reality, be highly phosphorylated. Thus the difficulty in labelling them with [^{32}P]orthophosphate could be because of slow rates of turnover of their phosphoryl groups. The main counter-argument to this is that treatment of ribosomes with alkaline phosphatase does not appear to cause any alteration in the position of migration of ribosomal proteins [50,146,143]. However, alkaline phosphatase might not be able to dephosphorylate these sites.

There are some reports of more extensive phosphorylation of proteins which did not correspond to known ribosomal components. Thus Schiffmann and Horak observed two additional phosphoproteins from HeLa cell ribosomes and designated them S32 and L42 [165]. However, these proteins remained at or near the origin during two-dimensional gel electrophoresis, and they may merely represent aggregated or precipitated states of other proteins. Zinker and Warner encountered a similar difficulty in two-dimensional gel analysis of their protein, P1 [17].

226

In contrast, the new phosphorylated components which appeared after infection of L cells with vesicular stomatitis virus were clearly resolved in two-dimensional gels [66]. However, the radioactive orthophosphate did not comigrate with any known ribosomal protein, and indeed the amount of protein migrating in this position was so low that it gave no visible stain. Thus the radioactive spots would appear to represent either extremely small amounts of phosphorylated ribosomal proteins which migrate at positions distinct from their parent proteins, or trace contamination with non-ribosomal proteins. The latter might possibly be histones which we have occasionally found to contaminate ribosomes [202].

The situation regarding additional ribosomal phosphoproteins is more complicated in yeast, the two-dimensional gel pattern of which is difficult to compare with that of higher organisms. Apart from one report claiming a total of fifteen phosphoproteins [16], most studies in this organism revealed two or three phosphoproteins additional to the acidic proteins and the presumed equivalents of S6 [17,18,55,91] (although in two of these reports the additional phosphoproteins were only slightly labelled [18,55]). It may be that in yeast, the acidic and S6-like proteins are the most phosphorylated but that phosphorylations similar to those of S2, S3 and L14 occur more readily than in mammalian cells.

5. Phosphorylation of ribosomal proteins by protein kinases in vitro

There have been a large number of reports of the phosphorylation of ribosomes in vitro by protein kinases from a wide range of eukaryotic cells. In some studies the protein kinase preparations used were derived from the cytosol fraction of the same or other cells [5,6,104,135,65,16,108,18,144, 165–181], in others the source of enzyme was endogenous protein kinase activity associated with, but extractable from the ribosomes themselves [3,4,138,15,171,18,182–195]. In some cases the phosphorylation was stimulated by cyclic AMP and in one case by cyclic GMP [178], whereas in others it was not. In many of these studies the phosphorylated ribosomal proteins were not characterised in any way and it is possible that some of the proteins phosphorylated were non-ribosomal contaminants (cf. ref. [196] and [197]).

It is not intended to review this work is detail here but largely to confine discussion to the kinases responsible for phosphorylating the major ribosomal phosphoproteins of intact cells. It will therefore be possible to consider only those studies in which the phosphorylated ribosomal proteins were properly identified. In one of the earliest and most thorough studies of phosphorylation of rat liver ribosomes by exogenous protein kinase over a dozen proteins were found to be phosphorylated [5,135], a result that can now be seen to demonstrate that the pattern of phosphorylation in vitro may not reflect that in the intact cell. Other studies also revealed a different pattern of phosphorylation in vitro from that now known to obtain in situ [168,170,104, 173,65,16,18], the most extreme example of this being the phosphorylation of E.coli ribosomal proteins, which are not phosphorylated in situ [45,46,41].

More recently studies have been performed on the phosphorylation of ribosomal proteins by purified protein kinases with analysis on two-dimensional polyacrylamide gels. Our own results (McGarvey and Leader, unpublished) are in general agreement with others that the acidic phospho-proteins of the 60 S ribosomal subunit are the sole ribosomal substrate for a cyclic AMP-independent protein kinase that can also phosphorylate casein and use both ATP and GTP as phosphoryl donors [175,176,144,180] (type III in the classification of Traugh and Traut [198]). It is assumed, but not proved by phosphopeptide isolation, that this is the kinase phosphorylating these proteins in vivo. This kinase appears to be distinct from another protein kinase that utilises GTP but which phosphorylates the basic protein, S17, of the 40 S ribosomal subunit in vitro [104].

Ribosomal protein S6 can also be phosphorylated in vitro using cAMP-dependent protein kinase (Type I in the aforementioned classification), but the same protein kinase phosphorylates proteins on the 60 S subunit as well [65]. It is not clear whether this means that the kinase in question is not involved in phosphorylation of ribosomal proteins in vivo, or whether it is so involved and its phosphorylation of 60 S ribosomal proteins in vitro is an artefact. In this respect it would be interesting to see characterised the endogenous protein kinase activity on rat liver ribosomes which is specific for ribosomal protein S6 [192,199] (see Section 2.4). The characterisation of the protein kinase(s) responsible for the phosphorylation of S6 in situ is clearly necessary for an understanding of the control of the extent of phosphorylation of this protein. Identification of the protein kinase activities responsible for phosphorylating S3 and L14 would be interesting for similar reasons.

In contrast to the wealth of data on ribosomal phosphoprotein kinases, there are only a few reports of ribosomal phosphoprotein phosphatases. These relate exclusively to ribosomal protein S6 which has been dephospory-lated in rat liver ribosomes using a phosphoprotein phosphatase from skeletal

228

muscle [200,9], and in rabbit reticulocyte ribosomes using a rabbit reticulocyte enzyme [201,189]. More work is required to determine the relative importance of phosphoprotein kinases and phosphoprotein phosphatases in controlling the extent of phosphorylation of individual ribosomal proteins.

Acknowledgements

It is a pleasure to acknowledge the colleagues who have contributed to different aspects of the author's studies: Amanda Coia, Laila Fahmy, Iain Kennedy, Michael McGarvey and Andrew Rankine. Various parts of this work have been funded by grants to the author from the British Diabetic Association and the Medical Research Council, and by a Grant to Professor R.M.S. Smellie from the Cancer Research Campaign. I am grateful to Drs. S. Lastick, E. McConkey, W. Möller, K. Ogata, G. Thomas and I.G. Wool for providing me with access to manuscripts prior to their publication.

References

1. Kabat, D. (1970) Biochemistry, 9, 4160–4175.
2. Loeb, J.E. and Blat, C. (1970) FEBS Lett., 10, 105–108.
3. Kabat, D. (1971) Biochemistry, 10, 197–203.
4. Li, C.C. and Amos, H. (1971) Biochem. Biophys. Res. Commun., 45, 1398–1407.
5. Eil, C. and Wool, I.G. (1971) Biochem.Biophys. Res. Commun., 43, 1001–1009.
6. Walton, G.M., Gill, G.N., Abrass, I.B. and Garren, L.D. (1971) Proc. Natl. Acad. Sci. USA, 68, 880–884.
7. Krystosek, A., Bitte, L.F., Cawthon, L.M. and Kabat, D. (1974) in: Ribosomes (Nomura, M., Tissières, A. and Lengyel, P., eds.), pp. 855–870, Cold Spring Harbor, New York.
8. Krystosek, A., Bitte, L.F. and Kabat, D. (1974) in: Proceedings of the 3rd International Symposium on the Metabolic Interconversion of Enzymes (Fischer, E.H., Krebs, E.G., Neurath, H. and Stadtman, E.R., eds.), pp. 165–173, Springer-Verlag, Berlin.
9. Loeb, J.E., Pierre, M., Perlès, B. and Creuzet, C. (1975) in: Post-synthetic Modification of Macromolecules (Antoni, F. and Faragó, A., eds.), pp. 149–157, North-Holland, Amsterdam.
10. Wool, I.G. and Stöffler, G. (1974) in: Ribosomes (Nomura, M., Tissières, A. and Lengyel, P., eds.), pp. 417–460, Cold Spring Harbor, New York.
11. Wool, I.G. (1979) Ann. Rev. Biochem., 48, 719–754.
12. Traut, R.R., Howard, G.A. and Traugh, J.A. (1974) in: Proceedings of the 3rd International Symposium on the Metabolic Interconversion of Enzymes (Fischer, E.H., Krebs, E.G., Neurath, H. and Stadtman, E.R., eds.), pp. 155–164, Springer-Verlag, Berlin.
13. Taborsky, G. (1974) Adv. Protein Chem., 28, 1–210.
14. Trewavas, A. (1973) Plant Physiol., 51, 760–767.
15. Grankowski, N. and Gasior, E. (1975) Acta Biochim.Polon., 22, 45–56.
16. Becker-Ursic, D. and Davies, J. (1976) Biochemistry, 15, 2289–2296.
17. Zinker, S. and Warner, J.R. (1976) J. Biol. Chem., 251, 1799–1807.

17. Zinker, S. and Warner, J.R. (1976) J. Biol. Chem., 251, 1799–1807.
18. Hébert, J., Pierre, M. and Loeb, J.E. (1977) Eur. J. Biochem., 72, 167–174.
19. Van Agthoven, A.J., Maasen, J.A. and Möller, W. (1977) Biochem. Biophys. Res. Commun., 77, 989–998.
20. Van Agthoven, A.J., Kriek, J., Amons, R. and Möller, W. (1978) Eur. J. Biochem., 91, 553–556.
21. Bélanger, G., Bellemarie, G. and Lemieux, G. (1979) Biochem. Biophys. Res. Commun., 86, 862–868.
22. Kristiansen, K., Plesner, P. and Krüger, A. (1978) Eur. J. Biochem., 83, 395–403.
23. Kristiansen, K. and Krüger, A. (1978) Biochim. Biophys. Acta, 521, 435–451.
24. Kaerlein, M. and Horak. I. (1976) Nature, 259, 150–151.
25. Lastick, S.M., Nielsen, P.J. and McConkey, E.H. (1977) Molec. Gen. Genet., 152, 223–230.
26. Sanecka-Obacz, M. and Borowski, T. (1974) Acta Biochim. Polon., 21, 397–401.
27. Gressner, A.M. and Wool I.G. (1974) Biochem. Biophys. Res. Commun., 60, 1482–1490.
28. Bitte, L. and Kabat, D. (1972) J. Biol. Chem., 247, 5345–5350.
29. Rankine, A.D. and Leader, D.P. (1975) FEBS Lett., 52, 284–287.
30. Leader, D.P., Rankine, A.D. and Coia, A.A. (1976) Biochem. Biophys. Res. Commun., 71, 966–974.
31. Rupp, R.G., Humphrey, R.M. and Shaeffer, J.R. (1976) Biochim. Biophys. Acta, 418, 81–92.
32. Barden, N. and Labrie, F. (1973) Biochemistry, 12, 3096–3102.
33. Hill, A.M. and Trachewsky, D. (1974) J. Steroid Biochem., 5, 561–568.
34. Ashby, C.D. and Roberts, S. (1975) J. Biol. Chem., 250, 2546–2555.
35. Roos, B.A. (1973) Endocrinology, 93, 1287–1293.
36. Majumder, G.C. and Turkington, R.W. (1972) J. Biol. Chem., 247, 7207–7217.
37. Stahl, J., Böhm, H. and Bielka, H. (1974) Acta Biol. Med. Ger., 33, 667–676.
38. Jolicoeur, P., Lemay, A., Labrie, F. and Steiner, A.L. (1974) Exp. Cell Res., 89,. 231–240.
39. Gordon, J. (1971) Biochem. Biophys. Res. Commun., 44, 579–586.
40. Rahmsdorf, H.J., Herrlich, P., Pai, S.H., Schweiger, M. and Wittmann, H.G. (1973) Molec. Gen. Genet., 127, 259–271.
41. Issinger, O.G., Kiefer, M.C. and Traut, R.R. (1975) Eur. J. Biochem., 59, 137–143.
42. Kurek, E., Grankowski, N. and Gasior, E. (1972) Acta Microbiol. Polon. Ser. A, 4 (21) 171–176.
43. Kurek, E., Grankowski, N. and Gasior, E. (1972) Acta Microbiol. Polon. Ser. A, 4 (21) 177–183.
44. Rahmsdorf, H.J., Pai, S.H., Ponta, H., Herrlich, P., Roskoski, R., Schweiger, M. and Studier, F.W. (1974) Proc. Natl. Acad. Sci. USA, 71, 586–589.
45. Traugh, J.A. and Traut, R.R. (1972) Biochemistry, 11, 2503–2509.
46. Issinger, O.G. and Traut, R.R. (1974) Biochem. Biophys. Res. Commun., 59, 829–836.
47. Rankine, A.D., Leader, D.P. and Coia, A.A. (1977) Biochim. Biophys. Acta, 474, 293–307.
48. Kabat, D. (1972) J. Biol. Chem., 247, 5338–5344.
49. Pierre, M., Creuzet, C. and Loeb, J.E. (1974) FEBS Lett., 45, 88–91.
50. Gressner, A.M. and Wool, I.G. (1974) J. Biol. Chem., 249, 6917–6925.
51. Leader, D.P. and Wool, I.G. (1972) Biochim. Biophys. Acta, 262, 360–370.
52. Kaltschmidt, E. and Wittmann, H.G. (1970) Anal. Biochem., 36, 401–412.
53. Leader, D.P. and Coia, A.A. (1978) Biochem. J., 176, 569–572.
54. Leader, D.P. and Coia, A.A. (1978) Biochim. Biophys. Acta, 519, 213–223.
55. Kruiswijk, T., De Hey, J.T. and Planta, R.J. (1978) Biochem. J., 175, 213–219.
56. Correze, C., Pinell, P. and Nunez, J. (1972) FEBS Lett., 23, 87–91.
57. Bitte, L and Kabat, D. (1974) Meth. Enzymol., 30, 563–590.
58. Leader, D.P. and Coia, A.A. (1978) Biochim. Biophys. Acta, 519, 224–232.

230

59. Cawthon, M.L., Bitte, L.F., Krystosek, A. and Kabat, D. (1974) J. Biol. Chem., 249, 275–278.
60. McConkey, E.H., Bielka, H., Gordon, J., Lastick, S.M., Lin, A., Ogata, K., Reboud, J.P., Traugh, J.A., Traut, R.R., Warner, J.R., Welfle, H. and Wool, I.G. (1979) Molec. Gen. Genet., 169, 1–6.
61. Sherton, C.C. and Wool, I.G. (1972) J. Biol. Chem., 247, 4460–4467.
62. Madjar, J.J., Arpin, M., Buisson, M. and Reboud, J.P. (1979) Molec. Gen. Genet., 171, 121–134.
63. Leader, D.P. (1975) Biochem. J., 152, 373–378.
64. Treolar, M.A., Treolar, M.E. and Kisilevsky, R. (1977) J. Biol. Chem., 252, 6217–6221.
65. Traugh, J.A. and Porter, G.G. (1976) Biochemistry, 15, 610–616.
66. Marvaldi, J. and Lucas-Lenard, J. (1977) Biochemistry, 16, 4320–4327.
67. Roberts, S. and Ashby, C.D. (1978) J. Biol. Chem., 253, 288–296.
68. Rosnitschek, I., Traub, U. and Traub, P. (1978) Z. Physiol. Chem., 359, 593–600.
69. Kruppa, J. and Martini, O.H.W. (1978) Biochem. Biophys. Res. Commun., 85, 428–435.
70. Haselbacher, G.K., Humbel, R.E. and Thomas, G. (1979) FEBS Lett., 100, 185–190.
71. Blat, C. and Loeb, J.E. (1971) FEBS Lett., 18, 124–126.
72. Gressner, A.M. and Wool, I.G. (1976) J. Biol. Chem., 251, 1500–1504.
73. Gressner, A.M. and Wool, I.G. (1976) Nature, 259, 148–150.
74. Schubart, U.K., Shapiro, S., Fleischer, N. and Rosen, O.M. (1977) J. Biol. Chem., 252, 92–101.
75. Rankine, A.D. (1976) Ph.D. Thesis, University of Glasgow
76. Horak, I. and Koschel, K. (1977) FEBS Lett., 83, 68–70.
77. Leader, D.P. and Coia, A.A. (1978) FEBS Lett., 90, 270–274.
78. Lastick, S.M. and McConkey, E.H. (1978) in: ICN-UCLA Symposium on Molecular and Cellular Biology (Dirksen, E.R., Prescott, D.M. and Fox, C.F., eds.), Vol. 12, pp. 61–69.
79. Kaerlein, M. and Horak, I. (1978) Eur. J. Biochem., 90, 463–469.
80. Ziv, E. and Stratman, F.W. (1976) FEBS Lett., 68, 86–88.
81. Werenne, J., Laurel, R. and Laurel, A. (1973) Arch. Int. Physiol. Biochim., 81, 602.
82. Rankine, A.D. and Leader, D.P. (1975) Biochem. Soc. Trans., 3, 546.
83. Leader, D.P. and Rankine, A.D. (1976) in: Ribosomes and RNA Metabolism (Zelinka, J. and Balan, J., eds.), Vol. 2, pp. 319–326, Slovak Academy of Sciences, Bratislava.
84. Appleman, M.M. and Kemp, R.G. (1966) Biochem. Biophys. Res. Commun., 24, 564–568.
85. Wititsuwannakul, D. and Kim, K.H. (1977) Biochem. Biophys. Res. Commun., 76, 86–92.
86. Ziv, E., Wagner, M.J. and Stratman, F.W. (1978) FEBS Lett., 86, 219–224.
87. Rall, T.W. and Sutherland, F.W. (1962) J. Biol. Chem., 237, 1228–1232.
88. Gressner, A.M. and Greiling, H. (1977) FEBS Lett., 74, 77–81.
89. Gressner, A.M. and Greiling, H. (1978) Exp. Mol. Pathol., 28, 39–47.
90. Gressner, A.M. and Greiling, H. (1979) Biochem. Pharmacol., 27, 2495–2498.
91. Otaka, E. and Kobata, K. (1978) Molec. Gen. Genet., 162, 259–268.
92. Russell, W.C. and Blair, G.E. (1977) J. Gen. Virol., 34, 19–35.
93. Blair, G.E. and Horak, I. (1977) Biochem. Soc. Trans., 5, 660–661.
94. Carasco, L. and Smith, A.E. (1976) Nature, 264, 807–809.
95. Zeilig, C.E., Johnson, R.A., Sutherland, E.W. and Friedman, D.L. (1976) J. Cell. Biol., 71, 515–534.
96. Rudland, P.S., Seeley, M. and Seifert, W. (1974) Nature, 259, 148–150.
97. Anderson, W.M., Grundholm, A. and Sells, B.H. (1975) Biochem. Biophys. Res. Commun., 62, 669–676.
98. Scheinbuks, J., Sypherd, P.S. and Moldave, K. (1974) Biochem. Biophys. Res. Commun., 61, 322–328.
99. Tas, P.W.L. and Sells, B.H. (1978) Eur. J. Biochem., 92, 271–278.
100. MacManus, J.P., Franks, D.J., Youdale, T. and Braceland, B.M. (1972) Biochem. Biophys. Res. Commun., 49, 1201–1207.

101. Barela, T.D. and Kizer, D.E. (1974) Biochim. Biophys. Acta, 335, 218–225.
102. Hoffman, W.L. and Ilan, J. (1975) Mol. Biol. Rep., 2, 219–224.
103. Collatz, E., Wool, I.G., Lin, A. and Stöffler, G. (1976) J. Biol. Chem., 251, 4666–4672.
104. Ventamiglia, F.A. and Wool, I.G. (1974) Proc. Natl. Acad. Sci. USA, 71, 350–354.
105. Gross, B. and Westermann, P. (1976) Chem. Biol. Interact., 15, 309–317.
106. Kisilevsky, R., Weiler, L. and Treloar, M. (1978) J. Biol. Chem., 253, 7101–7108.
107. Kruiswijk, T., Planta, R.J. and Krop, J.M. (1978) Biochim. Biophys. Acta, 517, 378–389.
108. Cenatiempo, Y., Cozzone, A.J., Genot, A. and Reboud, J.P. (1977) FEBS Lett., 79, 165–169.
109. Welfle, H., Goerl, M. and Bielka, H. (1978) Molec. Gen. Genet., 163, 101–112.
110. Metspalu, A., Saarma, M., Villems, R., Ustav, M. and Lind, A. (1978) Eur. J. Biochem., 91, 73–81.
111. Bielka, H., Noll, F., Welfle, H., Westermann, P., Lutsch, G., Stahl, J., Thiese, H., Bommer, U.-A., Gross, B., Goerl, M. and Henkel, B. (1979) in Gene Functions (Rosenthal, S., Bielka, H., Contelle, Ch. and Zimmer, Ch., eds.) Proceedings of 12th FEBS Meeting, Vol. 51, pp. 387–399, Pergamon Press, Oxford.
112. Terao, K. and Ogata, K. (1979) J. Biochem., 86, 605–617.
113. Fischer, N., Stöffler, G. and Wool, I.G. (1978) J. Biol. Chem., 253, 7355–7360.
114. Khairallah, E.A. and Pitot, H.C. (1967) Biochem. Biophys. Res. Commun., 29, 269–274.
115. Adiga, P.R., Murthy, P.V.N. and McKenzie, J.M. (1971) Biochemistry, 10, 711–715.
116. Wu, J.M., Cheung, C.P. and Suhadolnik, R.J. (1978) J. Biol. Chem., 253, 8578–8582.
117. Varrone, S., Di Lauro, R., and Macchia, V. (1973) Arch. Biochem. Biophys., 157, 334–338.
118. Klaipongpan, A., Bloxham, D.P. and Akhtar, M. (1975) FEBS Lett., 58, 81–84.
119. Buchwald, I., Hackett, P.B., Egberts, E. and Traub, P. (1977) Mol. Biol. Rep., 3, 315–321.
120. Monier, D., Santhanam, K. and Wagle, S.R. (1972) Biochem. Biophys. Res. Commun., 46, 1881–1886.
121. Datta, A., De Haro, C., Sierra, J.M. and Ochoa, S. (1977) Proc. Natl. Acad. Sci. USA, 74, 1463–1467.
122. Kramer, G., Henderson, A.B., Pinphanichakarn, P., Wallis, M.H. and Hardesty, B. (1977) Proc. Natl. Acad. Sci. USA, 74, 1445–1449.
123. Yeung, D. and Oliver, I.T. (1968) Biochemistry, 7, 3231–3239.
124. Wicks, W.D., Kenney, F.T. and Lee, K.L. (1969) J. Biol. Chem., 244, 6008–6013.
125. Iynedjian, P.B. and Hanson, R.W. (1977) J. Biol. Chem., 252, 655–662.
126. Roper, M.D. and Wicks, W.D. (1978) Proc. Natl. Acad. Sci. USA 75, 140–148.
127. Pain, V.M. and Garlick, P.J. (1974) J. Biol. Chem., 249, 4510–4514.
128. Peavy, D.E., Taylor, J.M. and Jefferson, L.S. (1978) Proc. Natl. Acad. Sci. USA 75, 5879–5883.
129. Wool, I.G., Castles, J.J., Leader, D.P. and Fox, A. (1972) Handbook of Physiology, Sec. 7, Endocrinology, Vol. 1, pp. 385–394.
130. Olson, M.O.J., Prestayko, A.W., Jones, C.E. and Busch, H. (1974) J. Mol. Biol., 90, 161–168.
131. Abelson, H.T., Johnson, L.F., Penman, S. and Green, H. (1974) Cell, 1, 161–165.
132. Kristiansen, K. and Krüger, A. (1979) Exp. Cell Res., 118, 159–169.
133. Rudland, P.S., Weil, S. and Hunter, A.R. (1975) J. Mol. Biol., 96, 745–766.
134. Eil, C. and Wool, I.G. (1973) J. Biol. Chem., 248, 5130–5136.
135. Eil, C. and Wool, I.G. (1973) J. Biol. Chem., 248, 5122–5129.
136. Rankine, A.D. and Leader, D.P. (1976) Mol. Biol. Rep., 2, 525–528.
137. Tsurugi, K., Collatz, E., Todokoro, K., Ulbrich, N., Lightfoot, H.N. and Wool, I.G. (1978) J. Biol. Chem., 253, 946–955.
138. Martini, O.H.W. and Gould, H.J. (1973) Biochim. Biophys. Acta, 295, 621–629.
139. Leader, D.P. and Coia, A.A. (1977) Biochem. J., 162, 199–200.
140. Creusot, F., Delaunay, J. and Shapira, G. (1975) Biochimie, 57, 167–173.

141. Howard, G.A., Traugh, J.A., Croser, E.A. and Traut, R.R. (1975) J. Mol. Biol., 93, 391–404.
142. Terao, K. and Ogata, K. (1975) Biochim. Biophys. Acta, 402, 214–229.
143. Leader, D.P., Coia, A.A. and Fahmy, L.H. (1978) Biochem. Biophys. Res. Commun., 83, 50–58.
144. Horak, I. and Schiffman, D. (1977) Eur. J. Biochem., 79, 375–380.
145. Arpin, M., Madjar, J.J. and Reboud, J.P. (1978) Biochim. Biophys. Acta, 519, 537–541.
146. Houston, L.L. (1978) Biochem. Biophys. Res. Commun., 85, 131–139.
147. Reyes, R., Vázquez, D. and Ballesta, J.P.G. (1977) Eur. J. Biochem., 73, 25–31.
148. Terhorst, C., Möller, W., Laursen, R. and Wittmann-Liebold, B. (1973) Eur. J. Biochem., 34, 138–152.
149. Subramian, A.R. (1975) J. Mol. Biol., 95, 1–8.
150. Möller, W. (1974) in: Ribosomes (Nomura, M., Tissières, A. and Lengyel, P., eds.), pp. 711–731, Cold Spring Harbor, New York.
151. Stöffler, G., Wool I.G., Lin, A. and Rak, K.H. (1974) Proc. Natl. Acad. Sci. USA, 72, 4744–4748.
152. Howard, G.A., Smith, R.L. and Gordon, J. (1976) J. Mol. Biol., 106, 623–637.
153. Grasmuk, H., Nolan, R.D. and Drews, J. (1977) Eur. J. Biochem., 79, 93–102.
154. Hamel, E., Koka, M. and Nakamoto, T. (1972) J. Biol. Chem., 247, 805–814.
155. Richter, D. and Möller, W. (1974) in Lipmann Symposium (Richter, D., ed.), pp. 524–533, Walter de Gruyter, Berlin.
156. Möller, W., Slobin, L.I., Amons, R. and Richter, D. (1975) Proc. Natl. Acad. Sci. USA 72, 4744–4748.
157. Amons, R., Van Agthoven, A., Pluijms, W., Möller, W., Higo, K., Itoh, T. and Osawa, S. (1977) FEBS Lett., 81, 308–310.
158. Amons, R., Pluijms, W. and Möller, W. (1979) FEBS Lett., 104, 85–89.
159. Sherton, C.C. and Wool, I.G. (1974) J. Biol. Chem., 249, 2258–2267.
160. Tischendorf, G.W., Zeichardt, H. and Stöffler, G. (1975) Proc. Natl. Acad. Sci. USA 74, 4820–4824.
161. Kruiswijk, T., Planta, R.J. and Mager, W.H. (1978) Eur. J. Biochem., 83, 245–252.
162. Collatz, E., Ulbrich, N., Tsurugi, K., Lightfoot, H.N., MacKinlay, W., Lin, A. and Wool, I.G. (1977) J. Biol. Chem., 252, 9071–9080.
163. Westermann, P., Heumann, W., Bommer, U.A., Bielka, H., Nygard, O. and Hultin, T. (1979) FEBS Lett., 97, 101–104.
164. Noll, F., Bommer, U.A., Lutsch, G., Theise, H. and Bielka, H. (1978) FEBS Lett., 87, 129–131.
165. Schiffmann, D. and Horak, I. (1978) Eur. J. Biochem., 82, 91–95.
166. Yamamura, M., Inoue, Y., Shimomura, R. and Nishizuka, Y. (1972) Biochem. Biophys. Res. Commun., 46, 589–596.
167. Stahl, J., Welfle, H. and Bielka, H. (1972) FEBS Lett., 26, 233–236.
168. Walton, G.M. and Gill, G.N. (1973) Biochemistry, 12, 2604–2611.
169. Delaunay, J., Loeb, J.E., Pierre, M. and Schapira, G. (1973) Biochim. Biophys. Acta, 312, 147–151.
170. Traugh, J.A., Mumby, M. and Traut, R.R. (1973) Proc. Natl. Acad. Sci. USA, 70, 373–376.
171. Grankowski, N., Kudlicki, W. and Gasior, E. (1974) FEBS Lett., 47, 103–106.
172. Böhm, H. and Stahl, J. (1974) Acta Biol. Med. Ger., 32, 449–461.
173. Azhar, S. and Menon, K.M.J. (1975) Biochim. Biophys. Acta, 392, 64–74.
174. Grancharova, T.V., Getova, T.A. and Nikolov, T.K. (1976) Biochim. Biophys. Acta, 418, 397–403.
175. Kudlicki, W., Grankowski, N. and Gasior, E. (1976) Mol. Biol. Rep., 3, 121–129.
176. Issinger, O.G. (1977) Biochim. Biophys. Acta, 477, 185–189.
177. Issinger, O.G. (1977) Biochem. J., 16, 511–518.

178. Chihara-Nakashima, M., Hashimoto, E. and Nishizuka, Y. (1977) J. Biochem., 81, 1863–1867.
179. Horak, I. and Schiffmann, D. (1977) FEBS Lett., 82, 82–84.
180. Kudlicki, W., Grankowski, N. and Gasior, E. (1978) Eur. J. Biochem., 84, 493–498.
181. Duvernay, V.H. and Traugh, J.A. (1978) Biochemistry, 17, 2045–2049.
182. Fontana, J.A., Picciano, D. and Lovenberg, W. (1972) Biochem. Biophys. Res. Commun., 49, 1225–1232.
183. Jergil, B. (1972) Eur. J. Biochem., 28, 546–554.
184. Pavlovic-Hournac, M., Delbauffe, D., Virion, A. and Nunez, J. (1973) FEBS Lett., 33, 65–69.
185. Schmidt, M.J. and Sokoloff, L. (1973) J. Neurochem., 21, 1193–1205.
186. Keates, R.A.B. and Trewavas, A.J. (1974) Plant Physiol., 54, 95–99.
187. Kozyreff, V. and Devisscher, M. (1974) Biochim. Biophys. Acta, 362, 17–28.
188. Hardie, D.G. and Stansfield, D.A. (1977) Biochem. J., 164, 213–221.
189. Traugh, J.A. and Sharp, S.B. (1977) J. Biol. Chem., 252, 3738–3744.
190. Segawa, K., Yamaguchi, N. and Oda, K. (1977) J. Virol., 22, 679–693.
191. Schmitt, M., Kempf, J. and Quirin-Stricker, C. (1977) Biochim. Biophys. Res. Commun., 481, 438–449.
192. Genot, A., Reboud, J.P., Cenatiempo, Y. and Cozzone, A.J. (1978) FEBS Lett., 86, 103–107.
193. Cenatiempo, Y., Cozzone, A.J., Genot, A. and Reboud, J.P., (1978) Biochimie, 60, 813–816.
194. Segawa, K., Oda, K., Yuasa, Y., Shiroki, K. and Shimojo, H. (1978) J. Virol., 27, 800–808.
195. Tershak, D.R. (1978) Biochem. Biophys. Res. Commun. 80, 283–289.
196. Quirin-Stricker, C., Schmitt, M., Egly, J.M. and Kempf, J. (1976) Eur. J. Biochem., 62, 199–209.
197. Quirin-Stricker, C. and Schmitt, M. (1977) Biochim. Biophys. Acta, 477, 414–426.
198. Traugh, J.A. and Traut, R.R. (1974) J. Biol. Chem., 249, 1207–1212.
199. Genot, A., Reboud, J.P., Cenatiempo, Y. and Cozzone, A.J. (1979) FEBS Lett., 99, 261–264.
200. Perlès, B. and Loeb, J.E. (1974) Biochimie, 56, 1007–1010.
201. Lightfoot, H.N., Mumby, M. and Traugh, J.A. (1975) Biochem. Biophys. Res. Commun., 66, 1141–1146.
202. Leader, D.P. (1980) Biochem. J., 182, 241–245.
203. Kennedy, I.M. and Leader, D.P. (1980) Mol. Biol. Rep., in press

Modification of histone H1 by reversible phosphorylation and its relation to chromosome condensation and mitosis

HARRY R. MATTHEWS

1. Introduction

Histone H1 is a key component of chromatin. It stabilises the higher order packing of chromosome sub-units (nucleosomes) and probably exerts a controlling influence over the packing density in specific regions of the chromatin. It also appears to be involved in the massive condensation of chromatin into metaphase chromosomes that occurs in mitosis. Both these functions of H1 are probably controlled by reversible phosphorylation of H1. It has also been proposed that a reversible phosphorylation of H1 is involved in the assembly of nucleosomes on newly synthesised DNA. Thus, although H1 histone is not an enzyme in the usual sense of the word its function is regulated by reversible phosphorylation and so it falls into the context of this volume.

It is now widely accepted that chromatin is organised into sub-units, called nucleosomes, each of which contains a continuous 200 base-pair length of DNA and nine histone molecules. Eight of these histones, $(H3 \cdot H4)_2$. $H2A_2 \cdot H2B_2$, are bound together by hydrophobic interactions to form a core which is closely associated with about 145 base pairs of DNA. This histone·DNA complex is a well-defined, universal, structure which has already been studied in detail by a variety of methods (reviewed in [1]). Crystals of these complexes have been prepared and X-ray diffraction studies of these and larger crystals are expected to reveal their structure in detail. Neutron scattering and other studies have already shown that in solution the core is a flat disc of protein 5.5 nm thick and 7.0 nm in diameter with DNA wound round the edge to give an 11.0 nm diameter disc of DNA·protein

Cohen (ed.) Recently discovered systems of enzyme regulation by reversible phosphorylation which is called the core particle. A similar structure is found in the crystals. The positions of the remaining 55 base pairs of DNA (called linker DNA) and one histone molecule, H1, are not known and may depend on the higher order packing of the nucleosomes. H1 is probably bound directly to the linker DNA since H1 is lost when the linker DNA is digested by nuclease. The linker DNA may also be bound to core particle histones but this is not yet clear. Although the length of DNA in the core particle is constant the length of the linker DNA varies substantially between species, between cell types and within a single cell type [2,3].

A single metaphase chromosome may contain of the order of 0.1 m of DNA with its associated proteins in the form of nucleosomes (e.g. [4]). The nucleosomes are packed together in a compact structure that can form the metaphase chromosome. It has been proposed that this structure is organised round a protein matrix or scaffold which can be seen by electron microscopy of histone depleted chromosomes. The protein scaffold retains the shape of the metaphase chromosome and the DNA is partly released to form loops of length 30 000 to 90 000 base pairs [5,6]. Domains of comparable lengths have also been deduced from restriction nuclease digestion studies of chromatin [7,8]. It is assumed that in the presence of histones many core particles are formed in each loop and these interact with one another and with H1 histone and linker DNA to form the compact metaphase chromosome. These interactions can be modified either to cause the metaphase chromosome to disperse or to re-form by condensation of dispersed chromatin. The packing of core particles give rise to a higher order structure that has been called the quaternary structure [9] since the coiling of the double helix of DNA into the nucleosomes is regarded as the tertiary structure. Quaternary structures can be generated in chromatin solutions in the presence of Mg^{2+} ions and studied by electron microscopy, neutron diffraction or nuclease digestion. Electron microscopy [10] and neutron diffraction [11,12] both suggest a regular coil of nucleosomes. The coil or 'solenoid' can be visualised directly by electron microscopy and at the low chromatin concentrations used H1 histone was required in order to generate the quaternary coil. The pitch of the coil was 11 nm as if the nucleosomes in adjacent turns were very close to each other. Neutron diffraction of similarly structured chromatin showed a 10 nm diffraction maximum that was semi-meridionally oriented so that the diffraction was in the form of a cross of semi-angle 9°. The interpretation of this cross pattern is that the quaternary structure in chromatin consists of a flat helix of nucleosomes of pitch 10 nm with about 6 nucleosomes per turn. At the high chromatin concentration used to prepare samples for neutron diffraction the quaternary structure forms even in the absence of H1, suggesting in conjunction with the data from electron microscopy that the role of H1 is to stabilise the quaternary structure.

An alternative suggestion for the quaternary structure comes from different electron microscope studies and nuclease digestion experiments [13,14]. The electron micrographs show 'super beads', clusters of about 8 nucleosomes and the nuclease digestion studies show a tendency to generate DNA fragments of length corresponding to 8 or 16 or 24 nucleosomes. The relationship between these two structures, coil and superbead, has not yet been clarified.

Much earlier data supports the suggestion (above) that H1 is involved in chromosome packing (e.g. [15]). Sodium chloride induces a substantial contraction of chromatin gels which is completely absent in H1-depleted chromatin and a similar effect is seen in dilute solutions of chromatin where NaCl induces an H1-dependent turbidity in the solution. Since H1 interacts directly with DNA in chromatin, similar turbidity effects are obtained with H1·DNA complexes. The turbidity and gel contraction changes correlate with changes in the nuclear magnetic resonance spectrum of the H1 in the complex [16–18].

2. Sites of phosphorylation of histone H1

The amino acid sequence of histone H1 is less highly conserved than the sequences of the other histones and even within one cell type several minor variants of H1 can be resolved [19]. However, the main features of the molecule are universal. About 95 residues at the C-terminus contain most of the lysine, alanine and proline in the molecule and this region appears to have an extended structure which binds tightly to DNA. About 40 residues at the N-terminus also have an extended structure but their affinity for DNA is much less than that of the C-terminal region. Between these two regions there are about 77 residues whose sequence is highly conserved and which fold into a rigid globular structure which does not bind to DNA [20–26]. These three regions of H1 are summarised in Figure 1.

H1 can be phosphorylated by at least three different kinases or groups of kinases. The first protein kinases to be studied were extracted from rat liver and two separate activities were characterised. One kinase was the ubiquitous cyclic AMP dependent protein kinase and it was found to phosphorylate H1 mainly on serine-37. The other kinase was cyclic AMP independent and was found to phosphorylate H1 mainly on serine-106. The third group of histone kinases was found in rapidly dividing cells and is known as growth associated histone kinase, kinase-G [27–39]. This enzyme, or group of enzymes,

238

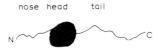

nose   head   tail

N   C

Figure 1. Diagrammatic representation of the structural regions of H1 histone showing the structured region from residue 40 to residue 117 and the N-terminal and C-terminal unstructured regions. The C-terminal region binds strongly to DNA.

phosphorylates multiple sites in the extended regions of H1. Four major sites have been identified: serine-180 and threonines-16, -136 and -153 [40]. Kinase-G from Ehrlich ascites cells has been purified to near homogeneity in low yield [32]. Kinase-G from a slime mould, *Physarum polycephalum* [41,42], has been partly purified and resolved into three components [42,33]. The different components have different substrate specificities and phosphorylate different sites on H1 in vitro. Their properties are discussed in more detail below.

The sites of phosphorylation of H1 are summarised in Figure 2. Notice that only one site, serine-106, occurs in the globular region of H1. Phosphorylation at this site has not been observed in vivo and it is assumed that the appropriate protein kinase involved acts on a different substrate in vivo. As might be expected phosphorylation of serine-106 has substantial effects in reducing the stability of the globular structure in the centre of H1. It also substantially reduces the ability of H1 to form aggregates with DNA [44,17].

Serine-37 is located at the boundary between the N-terminal region and the globular region of H1. In some H1 molecules serine-37 is replaced by alanine which is not phosphorylated [26,45]. Phosphorylation of serine-37 does occur in vivo. In particular it has been shown to occur as part of the response of liver cells to glucagon. The phosphorylation occurs specifically at serine-37 on about 2% of the H1 molecules [29,30,46]. This may be associated with

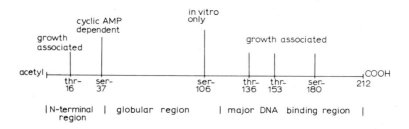

Figure 2. The major phosphorylation sites on H1 histone and their correlation with the structural regions of the molecule.

specific changes in transcription patterns of the chromatin although direct evidence for this lacking. Phosphorylation of serine-37 has rather little effect on the ability of H1 to form aggregates with DNA but this does not preclude other structural effects due to phosphorylation of serine-37 especially since it does reduce the strength of binding of H1 to DNA and it does reduce the distortion of DNA caused by its binding to H1 [17,44,47]. For example, the structure of transcribing chromatin is clearly different from that of non-transcribing chromatin although both are organised into a repeating sub-unit structure (e.g. [48]). The transition from non-transcribing to transcribing chromatin might involve phosphorylation of serine-37 on H1 but it probably involves non-histone proteins as well, for example HMG proteins [49,50]. Since serine-37 is not in the major DNA binding region of H1 its phosphorylation may affect interactions with non-histone proteins as well as with DNA.

The other four major sites so far identified are known as the growth associated sites and they are located in the terminal regions of H1, well away from the globular region. Phosphorylation of these sites also occurs in vivo but under quite different conditions from phosphorylation of serine-37. These sites are only phosphorylated in growing cells [51] where their phosphorylation is correlated with chromosome condensation [52, 53]. Phosphorylation of the growth associated sites substantially increases the ability of H1 to form aggregates with DNA [17,54].

3. Regulation of the function of H1 histone by phosphorylation

Phosphorylation of H1 at the growth associated sites is correlated with chromosome condensation in vivo. It has not yet been possible to reproduce chromosome condensation in vitro but studies with a highly simplified in vitro system have yielded very promising results. The system is based on an early observation that a form of chromatin contraction occurs in vitro when the sodium ion concentration in a chromatin gel is raised to about 0.15 M. The contraction does not occur in chromatin that has been depleted of H1 and nuclear magnetic resonance studies showed a correlation between the spectrum of H1 and chromatin contraction [16]. A similar phenomenon occurs in more dilute solutions of chromatin where it is conveniently followed by observing the turbidity of the chromatin solution as a function of sodium ion concentration. The turbidity behaviour and the nuclear magnetic resonance spectra of H1·DNA mixtures are similar to those from chromatin as would be expected since H1 probably interacts directly with DNA in chromatin. This is important because, until very recently, there was no method for reconstituting chromatin that gave H1 bound in a fashion resembling that in vivo. H1 could be removed and replaced, but during the process the core particles 'slid' along the DNA and micrococcal nuclease digestion did not generate the normal 'ladder' of DNA bands. However, methods for removing H1 without core particle movement were recently reported ([55], and Staynov, personal communication) and it is likely that the experiments described below will be repeated with chromatin.

Figure 3 shows the turbidity of a chromatin solution and of an H1·DNA mixture as functions of sodium ion concentration. Nuclear magnetic resonance spectra of similar samples imply that in water H1 is bound to DNA; in intermediate sodium ion concentrations H1 is more firmly bound to DNA; and that at high sodium ion concentrations it is bound more loosely and is completely released in about 0.5 M NaCl. The turbidity data show that both chromatin and H1·DNA mixtures are aggregated at intermediate sodium ion concentrations but disaggregated at low sodium ion concentration where H1 is still bound and at high sodium ion concentration where H1 is released from the DNA. These effects are not specific for sodium ions; other monovalent

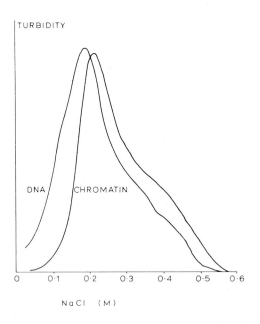

Figure 3. Turbidity as a function of sodium ion concentration for H1·DNA and for chromatin. The turbidity is caused by aggregation which in turn is caused by H1. The pattern of aggregation is similar for DNA and for chromatin. DNA alone or H1-depleted chromatin do not aggregate under these conditions.

and divalent anions show similar effects although much lower concentrations of divalent ions are required [23].

The aggregation process is affected by phosphorylation of H1 [17]. Phosphorylation of serine-106, which destabilises the structure of the globular region of H1, reduces the ability of H1 to aggregate DNA. Phosphorylation of serine-37 alone has a negligibly small effect on the ability of H1 to aggregate DNA except at high sodium ion concentration. However, phosphorylation at the growth associated sites increases the ability of H1 to aggregate DNA. The differences are summarised in Figure 4.

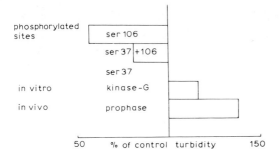

Figure 4. Ability of H1 to aggregate DNA as a function of phosphorylation of H1; summary of turbidity experiments. Phosphorylation of serine-106 reduces the ability of H1 to aggregate DNA whereas phosphorylation of the growth associated sites increases the ability of H1 to aggregate DNA. Even larger differences can be demonstrated by choosing a specific sodium ion concentration at which to make the comparison [54,17].

The experiments described so far used in vitro phosphorylated H1 molecules and the maximum degree of phosphorylation achieved was 1.0 phosphates per molecule for serine-37 or serine-106 and 2.2 phosphatases per molecule for the growth associated sites. Phosphorylation at both serine-37 and serine-106 gave 2.0 phosphates per molecule and a reduced ability of H1 to aggregate DNA showing that the effect of phosphorylation at the growth associated sites is specific for these sites and not a general effect of phosphorylation.

Recently, these studies have been extended to in vivo phosphorylated H1 [54] which was obtained using the naturally synchronous cell cycle in *Physarum polycephalum*. *Physarum* nuclei contain five major histone species corresponding to the five species found in mammalian cells [52,54,56–61]. H1 in *Physarum* has a slightly higher molecular weight than mammalian H1, 23 000–25 000 compared with 21 000–23 000 for mammalian H1. *Physarum* H1 is also slightly less basic, with 23% lysine + arginine compared with 28%

242

for calf thymus H1 [62], and has a higher proportion of arginine, 4% compared with 1.7% for calf thymus H1. These differences fall within the variations normally found for H1 from different species. Recent data from chymotrypsin digestion of *Physarum* H1 are consistent with the N-terminal half of the molecule, including the globular region, being conserved while the C-terminal half contains the variations [63]. In spite of these detail diffe-rences, *Physarum* H1 has an ability to aggregate DNA that is very similar to the ability of calf H1 [54]. Consequently, *Physarum* provides an excellent source of H1 isolated from precisely defined stages of the cell cycle without the use of inhibitors.

H1 was isolated and purified from naturally synchronous plasmodia of *Physarum* in either S phase or prophase of the cell cycle. Chromosome condensation occurs during prophase. S phase H1 has a low phosphate content, probably one phosphate per molecule, while prophase H1 has a 6-fold higher phosphate content [52]. Figure 5 shows their ability to aggregate DNA [54]. Prophase H1 is much more effective than S phase H1, implying that phosphorylation of H1 at the growth associated sites contributes to the process of chromosome condensation. Figure 4 shows that the in vivo phos-phorylated H1 is even better at aggregating DNA than the in vitro phosphory-lated H1, probably due to the greater number of phosphates per molecule for the in vivo phosphorylated H1.

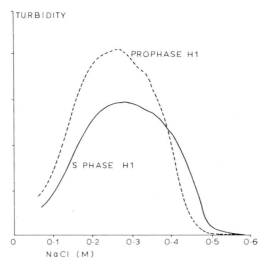

Figure 5. Turbidity as a function of sodium ion concentration for H1·DNA. H1 isolated from late prophase nuclei (high phosphate content) is more effective at aggregating DNA than H1 isolated from S phase nuclei (low phosphate content) except at high sodium ion concentration where the highly phosphorylated H1 is released from the DNA.

Phosphorylation of serine-37 has very little effect on the turbidity of H1·DNA mixtures at sodium ion concentrations up to 0.4 M which suggests that it has a different function from phosphorylation of the growth associated sites. However, phosphorylation of serine-37 does reduce the strength of binding of H1 to DNA as shown by nuclear magnetic resonance [44], circular dichroism [47] and turbidity [17]. Watson and Langan [64] reported that 'chromatin' reconstituted with H1 phosphorylated at serine-37 was a better template for RNA polymerase than 'chromatin' reconstituted with control H1. The fidelity of reconstitution was probably poor so the result may simply reflect the weaker binding of phosphorylated H1 to DNA. However, in vivo studies also show a possible correlation between transcription and phosphorylation of H1 at serine-37 [29,30].

It has been suggested that phosphorylation of H1 is involved in chromosome replication, in S phase of the cell cycle [65,51,66]. Although the major H1 phosphorylation event of the cell cycle occurs in prophase there is a significant phosphorylation of H1 during S phase (cf. [67]). There is still controversy over whether the sites involved are the same as those involved in prophase, or a sub-set of the prophase sites or uniques sites. However, Ajiro et al. [68] have recently presented convincing evidence for S phase specific turnover of phosphate at a particular, unidentified, site.

4. Phosphorylation of H1 histone in vivo

Early studies of H1 phosphorylation were dogged by technical problems [69–72] until Langan presented unequivocal evidence for phosphorylation of serine-37 in H1 [46]. This was subsequently shown to occur in whole perfused liver as a consequence of the action of glucagon [30]. These and other experiments led to the hypothesis that phosphorylation of serine-37 is involved in stimulating transcription of specific genes [35]. Detailed confirmation of this hypothesis has not yet been obtained but it is clearly consistent with the data so far available.

The amount of phosphorylation that occurs under these conditions is small, about 1 phosphate per 50 molecules of H1, which is consistent with specific areas of chromatin being affected. A much greater degree of phosphorylation occurs in growing cells where it has been shown that the average phosphate content of H1 in asynchronous cell populations is proportional to the cell growth rate [51]. It was originally speculated that this phosphorylation event was associated with DNA synthesis but experiments with colchicine arrested metaphase cells showed a high phosphate content in colchicine arrested chromosomes [37,38] and experiments using the naturally synchronous cell cycle in *Physarum* showed that the H1 phosphate content increases mostly in late G2 phase with a maximum in prophase and a very rapid fall in late mitosis as S phase begins [52]. Five years later, experiments with synchronised CHO cells showed that similar increases occurred in G2 phase and prophase of these mammalian cells with the very rapid fall occurring in late mitosis as G1 phase begins [53].

Minor differences in the detailed timing of the maximum phosphate content between *Physarum* and CHO cells were observed but it is not known if this was just a consequence of the drastic synchronisation procedures used with the CHO cells. It is not due to the action of phosphatase during isolation of H1 as has been claimed [73].

A detailed comparison between H1 phosphorylation in *Physarum* and in CHO cells can be made if the data are transformed to similar units. This is a matter of some importance because *Physarum* has some crucial experimental advantages for cell cycle studies, especially in G2 phase, and it is necessary to decide how far the data obtained in *Physarum* can be extrapolated to mammalian cells. The length of G2 phase is similar in both *Physarum* and cultured CHO cells but the complete synchrony that occurs naturally in *Physarum* cannot be obtained in CHO cell cultures. For comparison purposes an imaginary mixture of *Physarum* cultures with the same composition as the CHO cell culture was constructed for each time point. This procedure does not allow for side effects of the synchronisation processes, of course. The phosphate content of H1 in the imaginary culture was calculated from data on the phosphate content of *Physarum* H1 through the cell cycle, based on radio-activity measurements [52]. The phosphate content of CHO cell H1 was calculated from published gel electrophoresis profiles [53]. Despite some uncertainties in both methods, comparable values were obtained as shown in Figure 6. It is immediately apparent from Figure 6 that both *Physarum* and CHO cells have substantially similar patterns of H1 phosphorylation which supports the suggestion that H1 phosphorylation is an important part of a universal process, namely chromosome condensation. The detailed differences in phosphatase content may be due to the use of Colcemid and hydroxy urea in the CHO cell cultures. These data also show why H1 phosphate content increases as cell growth rate increases [51]. The time occupied by G2 phase is substantially independent of growth rate whereas the time of G1 phase varies, to give variable growth rate. Consequently G2 phase occupies a larger part of the cell cycle in rapidly growing cells, thus giving a higher average phosphate content in a rapidly growing culture.

The original experiments with *Physarum* showed a precise correlation of H1

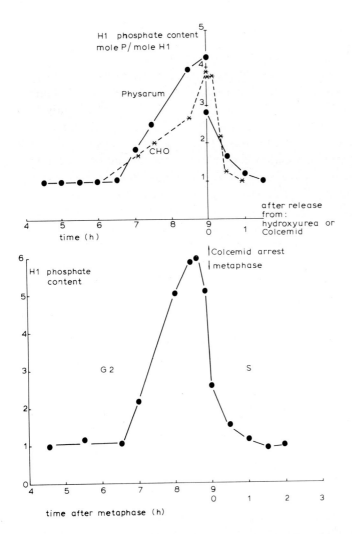

Figure 6. (a) Phosphate content of H1 as a function of time after release from hydroxyurea block in the presence of Colcemid and after removal of Colcemid for CHO cells. For comparison, the phosphate content of H1 in *Physarum* has been calculated for a similar, imaginary, experiment. (b) Phosphate content of *Physarum* H1 at precisely defined stages of the naturally synchronous nuclear division cycle showing the maximum H1 phosphate content in late prophase in these unperturbed nuclei.

phosphate content with the cell cycle processes leading to the formation of metaphase chromosomes and it was proposed, based on the detailed correlations, 'that phosphorylation of H1 is involved in chromosome condensation leading to mitosis' and '... is the initial mechanism of chromosome condensation' [52]. This view, that H1 phosphorylation is involved in initiation of chromosome condensation rather than maintenance of the mature metaphase chromosome structure has been challenged [73] but the original view is supported by additional data. In HeLa cells $ZnCl_2$ inhibits histone phosphatase but in the presence of $ZnCl_2$ (and consequently high histone phosphate content) the metaphase chromosomes de-condense, at least partially, as the cells move into G1 phase [74]. The results suggest that a high H1 phosphate content is not the major factor maintaining the metaphase chromosome structure. Data on histone phosphorylation in *Tetrahymena pyriformis* also support the view that high H1 phosphate is not required for maintenance of the condensed state of chromatin but that it is required for the initial stages of chromosome condensation [75]. Firstly, *Tetrahymena* micronuclei contain condensed chromatin but no H1 [76,77]. This may be analogous to the situation in nucleated erythrocytes where H5 replaces most of the H1 and to the situation in fish testes where protamines replace histones. In all these cases the chromatin is permanently condensed and so there is no longer a need to initiate chromosome condensation. Secondly, *Tetrahymena* macronuclei have phosphorylated H1 but divide amitotically [76,77]. Amitotic division in *Tetrahymena* does require the initial stages of chromosome condensation [78,79], although they are less marked in micronucleate strains, so the data are consistent with the above proposal that H1 phosphorylation initiates chromosome condensation.

The H1 phosphate content in *Physarum* is falling at metaphase [52], unlike the situation in metaphase arrested cells where the H1 phosphate content appears to reach a maximum. The very high phosphate content observed in metaphase arrested cells may be partly a consequence of metaphase arrest by colchicine, particularly as the chromosomes of such cells condense even further than those of normal cells. In this respect, the unperturbed cell cycle of *Physarum* provides a better system for cell cycle studies and in this organism the maximum phosphate content occurs in prophase, immediately before metaphase. This is consistent only with H1 phosphorylation initiating condensation. Further support for this view comes from studies of kinase activity in the cell cycle.

5. Growth associated histone kinase

Three kinases, or groups of kinases, associated with chromatin have already been mentioned. One is cyclic AMP dependent and is the common protein kinase also found in the cytoplasm. The second is cyclic AMP independent, as studied, but does not appear to phosphorylate H1 in vivo. The third, known as kinase-G is to be discussed here. It phosphorylates H1 at the growth associated sites and is probably responsible for the major H1 phosphorylation in growing cells.

It is hard to prove that a kinase is cyclic AMP independent. However, cyclic AMP dependence of kinase-G has never been reported and Bertulis [80,81] has looked at the problem carefully using crude nuclear extracts from *Physarum* as the source of kinase-G activity. A putative repressor sub-unit originally located in the nucleus would be expected to appear in the extract (the extract was prepared by sonicating nuclei in 0.5 M NaCl). The use of unfractionated extract reduces the chance of losing such a putative repressor sub-unit during purification but raises the possibility of other artefacts. In particular, the extract may contain cyclic AMP degrading activity (such as a phosphodiesterase) and/or a strong cyclic AMP binding activity, unconnected with the kinase. These possible artefacts were ruled out by measuring cyclic AMP concentration in the extract (using the competitive binding assay [82]) both as a function of added cyclic AMP and as a function of time. Figure 7 shows that cyclic AMP is neither strongly bound nor degraded under the conditions of kinase assay. Kinase-G is cyclic AMP independent under these

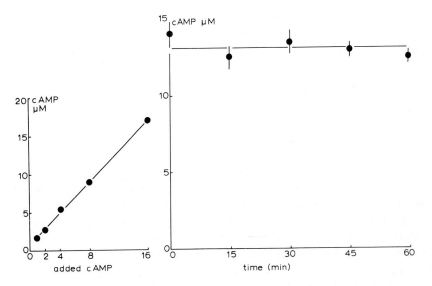

Figure 7. (a) Addition of cyclic AMP to the kinase-G extract. No loss of cyclic AMP occurs at zero time. (b) Incubation of cyclic AMP in the kinase-G extract. No loss of cyclic AMP occurs during the normal incubation time for kinase-G activity. Kinase-G activity is independent of cyclic AMP concentration under these conditions.

conditions. Further confirmation was provided by the finding that a heat stable inhibitor of cyclic AMP dependent kinases did not inhibit kinase-G. (The inhibitor was kindly provided by Dr. P. Cohen.) Partly and fully purified kinase-G is also cyclic AMP independent [37,41,32].

Langan has shown that the sites of H1 that are phosphorylated in vivo in rapidly growing Ehrlich ascites cells are similar to those phosphorylated in vitro by kinase-G from the same cells [41]. Preliminary results show that kinase-G from *Physarum* also phosphorylates these sites in vitro (Langan and Chambers, personal communication). The identity of sites supports the earlier suggestion from in vivo studies that the kinase-G activity, assayed in nuclear extracts, is the activity responsible for phosphorylating H1 in vivo in rapidly dividing cells [33,34].

Kinase-G activity has been measured through the naturally synchronous nuclear division cycle in *Physarum* and a 15-fold increase in activity was found in late G2 phase compared with S phase [34]. The increase just preceded the increase in H1 phosphate content implying that H1 phosphorylation was a consequence of the kinase-G activity increase and that initiation of chromosome condensation was then a consequence of H1 phosphorylation [34]. The authors were actually able to go further than this because the pattern of kinase-G activity correlated excellently with previous experiments using plasmodial fusion (e.g. [83,84]) or heat shock [85] or other methods (e.g. [86]) to perturb the time of mitosis in *Physarum*. The correlations led to the proposal that the increase in kinase-G activity is the major event in G2 phase controlling the initiation of mitosis. As a direct test of this proposal Inglis et al. [67,87] added kinase-G (partly purified from Ehrlich ascites chromatin) to growing plasmodia of *Physarum*. They found that the time of mitosis was advanced if kinase-G was added while endogenous kinase-G activity was increasing but not if kinase-G were added at other times nor if various control materials were added. Physarum kinase-G cross reacts immunologically with Ehrlich ascites kinase-G [88] and it has been shown recently that *Physarum* will take up a modified H3 histone from the growth medium and use it in chromatin (Prior and Cantor, personal communication). So it is reasonable to expect that kinase-G would also be taken up and used. This is not actually proved nor is it proved that advancement of mitosis was not due to another component of the kinase-G preparation, but nevertheless their series of experiments provides powerful support for the original proposal that increase of kinase-G activity controls the initiation of chromosome condensation and mitosis, in G2 phase.

Kinase-G, or a component of it, has been extracted from Ehrlich ascites chromatin and purified to near homogeneity [32]. The yield was too low for

hydration. The deuterium labelled kinase-G activity gave three bands at the same density as the unlabelled kinase-G, representing old enzyme, and 3 bands shifted by $+4$ to $+5$ mg/ml in the density gradient, representing labelled enzyme synthesised after transfer to growth medium which contained deuterated amino acids. Each labelled band was approximately the same size as its unlabelled partner on the same gradient showing that synthesis only gave a doubling of kinase-G activity, in line with total protein synthesis. The experiment thus proved that the 15-fold increase in kinase-G activity in G2 phase is mostly due to activation or de-repression of pre-existing enzyme.

The mechanism of activation of this group of kinases, kinase-G, is unknown but it appears that phosphorylation of H1 by kinase-G produces an important change in the properties of H1 which leads to initiation of chromosome condensation and hence mitosis. Further work is needed, not only on the kinases but also on the mechanism by which H1 stabilises the quaternary structure of chromatin and then on the way in which phosphorylation of H1 at the growth associated sites either promotes the formation of more quaternary structure in unstructured regions or initiates the packing of the quaternary structure into the next stage of higher order structure, which would lead directly to the metaphase chromosome.

Acknowledgements

Some of the work described here was supported by research grants from the Science Research Council, the Cancer Research Campaign and NATO.

I am grateful to Dr. S. Corbett for help in preparing Figure 4 from the data in her thesis [54] and to Mr. T. Chambers for the data in Figure 9.

References

1. Nicolini, C.A. (ed.) (1979) Chromatin structure and function, Plenum, New York.
2. Compton, J.C., Bellard, M. and Chambon, P. (1976) Proc. Natl. Acad. Sci. USA, 73, 4382–4386.
3. Johnson, E.M., Littau, V.C., Allfrey, V.G., Matthews, H.R. and Bradbury, E.M. (1976) Nucleic Acids Res., 3, 3313–3329.
4. Vogt, V. and Braun, R. (1976) FEBS Lett., 64, 190–192.
5. Adolph, K.W., Cheng, S.M. and Laemmli, U.K. (1977) Cell, 12, 805–816.
6. Paulson, J.R. and Laemmli, U.K. (1977) Cell, 12, 817–828.
7. Igo-Kemenes, T., Greil, W. and Zachau, H.G.A. (1977) Nucleic Acids Res., 4, 3387–3394.
8. Igo-Kemenes, T. and Zachau, H.G. (1978) Cold Spring Har. Symp. Quant. Biol., XLII, 109–115.

252

9. Rattle, H.W.E., Kneale, G.G., Baldwin, J.P., Matthews, H.R., Crane-Robinson, C., Cary, P.D., Carpenter, B.G., Suau, P. and Bradbury, E.M. (1979) in: Chromatin Structure and Function (C.A. Nicolini, ed.), pp. 451–514, Plenum, New York.
10. Finch, J.T., and Klug, A. (1976) Proc. Natl. Acad. Sci. USA, 73, 1897–1901.
11. Carpenter, B.G., Baldwin, J.P. Bradbury, E.M. and Ibel, K. (1976) Nucleic Acids Res., 3, 1739–1746.
12. Suau, P., Bradbury, E.M., and Baldwin, J.P. (1979) Eur. J. Biochem., in press.
13. Renz, M., Nehls, P. and Hozier, J. (1977) Proc. Natl. Acad. Sci. USA, 74, 1879–1883.
14. Olins, A.L. (1978) Cold Spring Harbour Symp. Quant. Biol. XLII, 325.
15. Littau, V.C., Burdick, C.J., Allfrey, V.G. and Mirsky, A.E. (1965) Proc. Natl. Acad. Sci. USA, 54, 1204–1209.
16. Bradbury, E.M., Carpenter, B.G. and Rattle, H.W.E. (1973) Nature, 241, 123–127.
17. Matthews, H.R. and Bradbury, E.M. (1978) Exp. Cell Res., 111, 343–351.
18. Bradbury, E.M., Danby, S.E., Rattle, H.W.E. and Giancotti, V. (1975) Eur. J. Biochem., 57, 97–105.
19. Kincade, J.M. and Cole, R.D. (1971) J. Biol. Chem., 241, 5790–5796.
20. Hartman, P.G., Chapman, G.E., Moss, T. and Bradbury, E.M. (1977) Eur. J. Biochem., 77, 45–51.
21. Chapman, G.E., Hartman, P.G., Cary, P.D., Bradbury, E.M. and Lee, D.R. (1978) Eur. J. Biochem., 86, 35–44.
22. Bradbury, E.M., Chapman, G.E., Danby, S.E., Hartman, P.G. and Riches, P.L. (1975) Eur. J. Biochem., 57, 521–528.
23. Danby, S.E. (1975) Ph.D. Thesis, Council for National Academic Awards, U.K.
24. Chapman, G.E., Hartman, P.G., and Bradbury, E.M. (1976) Eur. J. Biochem., 61, 69–75.
25. Bradbury, E.M., Cary, P.D., Chapman, G.E., Crane-Robinson, C., Danby, S.E. Rattle, H.W.E., Boublik, M., Palau, J. and Aviles, F.J. (1975) Eur. J. Biochem., 52, 605–613.
26. Rall, S.C. and Cole, R.D. (1971) J. Biol. Chem., 246, 7175–7190.
27. Langan, T.A. (1977) Methods Cell. Biol. 18, 127–142.
28. Langan, T.A., Hohman, P. (1975) in: Chromosomal Proteins and Their Role in the Regulation of Gene Expression (Stein, G.S. and Kleinsmith, L.J. ed.), pp. 113–125, Academic Press, New York.
29. Langan, T.A. (1969) Proc. Natl. Acad. Sci. USA, 64, 1276–1281.
30. Mallette, L.E., Neblett, M., Exton, J.H. and Langan, T.A. (1973) J. Biol. Chem., 248, 6289–6291.
31. Sherod, D., Johnson, G., Balhorn, R., Jackson, V., Chalkley, R. and Granner, D. (1975) Biochim. Biophys. Acta, 381, 337–347.
32. Schlepper, J. and Knippers, R. (1975) Eur. J. Biochem., 60, 209–220.
33. Hardie, D.G., Mattews, H.R. and Bradbury, E.M. (1976) Eur. J. Biochem., 66, 37–42.
34. Bradbury, E.M., Inglis, R.J., and Matthews, H.R. (1974) Nature (London), 247, 257–261.
35. Langan, T.A. (1971) Ann. N.Y. Acad. Sci., 185, 166–185.
36. Hohman, P., Tobey, R.A. and Gurley, L.R. (1975) Biochim. Biophys. Res. Commun., 63, 126–131.
37. Lake, R.S. and Salzman, N.P. (1972) Biochemistry, 11, 4817–4826.
38. Lake, R.S. (1973) J. Cell Biol., 58, 317–331.
39. Siebert, G., Ord, M.G. and Stocken, L.A. (1971) Biochem. J., 122, 721–725.
40. Langan, T.A. (1977) Methods Cell Biol., 18, 143–152.
41. Rusch, H.P. (1971) Adv. Cell Biol., 1, 297–327.
42. Mohberg, J. (1974) in: The Cell Nucleus (Busch, H., ed.) 1, 187–218.
43. Mitchelson, K., Chambers, T., Bradbury, E.M. and Matthews, H.R. (1978) FEBS Lett., 92, 339–342.
44. Rattle, H.W.E., Langan, T.A., Danby, S.E. and Bradbury, E.M. (1977) Eur. J. Biochem., 81, 499–503.
45. Cole, R.D. (1977) in: Molecular Biology of the Mammalian Genetic Apparatus (Ts'O, P.O.P. ed.), pp. 93–104, North-Holland, Amsterdam.

46. Langan, T.A. (1968) Science, 162, 579–580.
47. Adler, A.J., Schaffhausen, B., Langan, T.A. and Fasman, G.C. (1971) Biochemistry, 10, 909–913.
48. Matthews, H.R. (1977) Nature, 267, 203–204.
49. Goodwin, G.H., Walke, J.M. and Johns, E.W. (1978) in: The Cell Nucleus (Busch, H., ed.), Academic Press, New York.
50. Levy, W.B., Wong, N.C.W., and Dixon, G.H. (1977) Proc. Natl. Acad. Sci. USA, 74, 2810–2814.
51. Balhorn, R., Balhorn, M., Morris, H.P. and Chalkley, R. (1972) Cancer Res., 32, 1775–1784.
52. Bradbury, E.M., Inglis, R.J., Matthews, H.R. and Sarner, N. (1973) Eur. J. Biochem., 33, 131–139.
53. Gurley, L.R., D'Anna, J.A., Barham, S.S., Deaven, L.L. and Tobey, R.A. (1978) Eur. J. Biochem., 84, 1–16.
54. Corbett, S. (1979) Ph.D. Thesis. Council for National Academic Awards, U.K.
55. Zachau, H.G. (1979) in: Chromatin Structure and Function (Nicolini, C. ed.), Plenum Press, New York.
56. Mohberg, J. and Rusch, H.P. (1971) Exp. Cell Res., 66, 306–316.
57. Mohberg, J. and Rusch, H.P. (1969) Arch. Biochem. Biophys., 134, 577–585.
58. Mohberg, J. and Rusch, H.P. (1970) Arch. Biochem. Biophys., 138, 418–425.
59. Corbett, S., Miller, S., Robinson, V.J., Matthews, H.R. and Bradbury, E.M. (1977) Trans. Biochem. Soc., 5, 943–946.
60. Jochusch, B. and Walker, I.O. (1974) Eur. J. Biochem., 48, 417–425.
61. Johnson, E.M., Allfrey, V.C., Bradbury, E.M. and Matthews, H.R. (1978) Proc. Natl. Acad. Sci. USA, 75, 1116–1120.
62. Johns, E.W. (1976) in: Subnuclear Components; Preparation and Fractionation (Birnie, G.D., ed.), pp. 187–208, Butterworth, London.
63. Chambers, T. (1979) Ph.D. Thesis, Porstmouth Polytechnic.
64. Watson, G. and Langan, T.A. (1973) Fed. Proc., 32, 588.
65. Cross, M.E. and Ord, M.G. (1970) Biochem. J., 118, 191–197.
66. Marks, D.B., Paik, W.K. and Borun, T.W. (1973) J. Biol. Chem., 248, 5660–5667.
67. Bradbury, E.M., Inglis, R.J., Matthews, H.R. and Langan, T.A. (1974) Nature, 249, 553–556.
68. Dolby, T.W., Ajiro, K. and Borun, T. (1979) Biochemistry, 18, 1333–1343.
69. Stevely, W.S. and Stocken, L.A. (1968) Biochem. J., 110, 187–195.
70. Louie, A.J. and Dixon, G.H. (1973) Nature New Biol., 243, 164–168.
71. Sherod, D., Johnson, G. and Chalkley, R. (1970) Biochemistry, 9, 4611–4615.
72. Balhorn, R., Riecke, W.O. and Chalkley, R. (1971) Biochemistry, 10, 3952–3956.
73. Gurley, L.R., Tobey, R.A., Walters, R.A., Hilderbrand, C.E., Hohman, P.G. D'Anna, J.A., Barham, S.S. and Deaven, L.L. (1978) in: Cell Cycle Regulation (Jeter, J.R., Cameron, I.L., Padilla, G.M. and Zimmerman, A.M., ed.), Academic Press, New York.
74. Tanphaichitr, N., Moore, K.A., Granner, D. and Chalkley, R. (1976) J. Cell Biol., 69, 43–50.
75. Matthews, H.R. (1977) in: The Organisation and Expression of the Eukaryotic Genome (Bradbury, E.M. and Javaherian, K.,) pp. 67–80, Academic Press, London.
76. Gorovsky, M.A., Keevert, J.B. and Plager, G.L. (1974) J. Cell Biol., 61, 134–145.
77. Goroscky, M.A. and Keevert, J.B. (1975) Proc. Natl. Acad. Sci. USA,. 72, 2672–2676.
78. Zeuthen, E. and Resmussen, L. (1972) Res. Protozool., 4, 10–40.
79. Nilsson, J.R. (1970) J. Protozool., 17, 548–555.
80. Bertulis, M. (1978) Project Report, Portsmouth Polytechnic.
81. Matthews, H.R. (1979) in: Current Research on Physarum (Sachsenmaier, W., ed.), pp. 51–58, Publ. University of Innsbruch.
82. Cyclic AMP Assay, The Radiochemical Centre, Amersham, U.K.

83. Rusch, H.P., Sachsenmaier, W., Behrens, K. and Gruter, V. (1966) J. Cell. Biol., 31, 204–209.
84. Chin, B., Friedrich, P.D., Bernstein, I.A. (1973) J. Gen. Microbiol., 71, 93–101.
85. Brewer, E.N. and Rusch, H.P. (1965) Biochem. Biophys. Res. Commun., 21, 235–241.
86. Oppenheim, A. and Katzir, N. (1971) Exp. Cell Res., 68, 224–226.
87. Inglis, R.J., Langan, T.A., Matthews, H.R., Hardie, D.G. and Bradbury, E.M. (1976) Exp. Cell Res., 97, 418–425.
88. Zanker, K., Inglis, R.J., Matthews, H.R., Bradbury, E.M. (1977) Trans. Biochem. Soc., 5, 953–955.

CHAPTER 11

Protein phosphorylation and the co-ordinated control of intermediary metabolism

PHILIP COHEN

1. The molecular specificity of cyclic AMP-dependent protein kinase

The different regulatory systems described in this book have shown some striking similarities, and a number of general themes have started to emerge concerning the role of protein phosphorylation in the control of cellular function. It is the purpose of this concluding chapter to enlarge on some of these themes, which may be of general significance for future studies of enzyme regulation by reversible phosphorylation, since new examples of this phenomenon are appearing all the time.

A number of proteins have been described which are substrates for cyclic AMP-dependent protein kinase (pyruvate kinase, acetyl-CoA carboxylase, glycerol phosphate acyl transferase, cholesterol esterase, smooth muscle myosin kinase, troponin-I, phospholamban and histone H1). These recently discovered systems reinforce the hypothesis discussed in Chapter 1, that most, if not all, of the actions of cyclic AMP in mammalian tissues are mediated by phosphorylation reactions catalysed by cyclic AMP-dependent protein kinase. The increasing number of proteins that are being found to be substrates for cyclic AMP-dependent protein kinase therefore raises the question of the molecular specificity of this enzyme.

It is first of all important to stress that cyclic AMP-dependent protein kinase is a very specific enzyme which phosphorylates relatively few proteins at *significant* rates [1]. Even among substrate proteins, only one or two out of a large number of potentially available serine or threonine residues become phosphorylated. The determination of the amino acid sequences around some of the phosphorylation sites in recent years (Table 1) has demonstrated that

Cohen (ed.) Recently discovered systems of enzyme regulation by reversible phosphorylation
© *Elsevier/North-Holland Biomedical Press, 1980*

two adjacent basic amino acids, at least one of which is arginine, just N-terminal to the residue that is phosphorylated, is a feature common to all the best substrates for cyclic AMP-dependent protein kinase [1,2]. Work with synthetic peptides corresponding to some of these sites has confirmed that this structure feature is critical to specific substrate recognition (Chapter 2). The finding that the specificity of cyclic AMP-dependent protein kinase largely resides in the primary structure (or secondary structure determined by local primary structure) in the immediate vicinity of the phosphorylation site suggests a simple mechanism whereby proteins have evolved sensitivity to hormones. If, by mutation, a protein acquires two adjacent basic amino acids N-terminal to a serine (or occasionally a threonine) residue located on an accessible region on the surface of the protein, then that protein will become phosphorylated by cyclic AMP-dependent protein kinase in vivo. Should that phosphorylation then affect the function of the protein in a biologically useful way, the mutation will be selected for and spread through the population. It seems likely that the specificity of other protein kinases will also be found to reside in primary structures surrounding particular phosphorylation sites.

2. *Covalent versus non-covalent control of enzyme function*

An important general point brought out clearly by the analysis of pyruvate kinase (Chapter 2) and acetyl-CoA carboxylase (Chapter 3) as well as by glycogen phosphorylase and glycogen synthase (Chapter 1), is that enzymes regulated by phosphorylation-dephosphorylation are also controlled by a variety of allosteric effectors. Of course, this is not really surprising, since enzymes which occupy key positions in metabolism need to respond to alterations in the metabolic state of the cell as well as to physiological stimuli. However the extremely close interrelationship between control by phosphorylation-dephosphorylation and by allosteric effectors is worth re-emphasizing [3].

Allosteric effectors not only alter the activity of enzymes, but often affect the rate at which they are phosphorylated and dephosphorylated. Similarly, the effect of phosphorylation is often to change the K_m for a substrate, the K_a for an activator or the K_i for an inhibitor (or all three). In other words, whether the phosphorylation of an enzyme in vivo is ultimately expressed as a change in its activity, depends on the concentration of substrates and effector molecules in the cell, which have the power to either amplify or suppress the effects of phosphorylation. Phosphorylation–dephosphorylation should not therefore be thought of as a mechanism for switching the activity of an

Amino acid sequences at the phosphorylation sites of substrates for cAMP-dependent protein kinase

Substrate	Sequence[b]
Phosphorylase kinase (β-subunit)	100 Ala-Arg-Thr-Lys-Arg-Ser-Gly-Ser(P)-Val-Tyr-Glu-Pro-Leu-Lys Lys
Glycogen synthase (site Ia)	80 Lys-Arg-Ala-Ser(P) Arg
Pyruvate kinase, rat liver	35 Gly-Val-Leu-Arg-Arg-Ala-Ser(P)-Val-Ala-Glx-Leu
Pyruvate kinase, pig liver	? Leu-Arg-Arg-Ala-Ser(P)-Leu-Gly
Inhibitor-1	30 Ile-Arg-Arg-Arg-Pro-Thr(P)-Pro-Ala-Thr
Phosphorylase kinase (α-subunit)	20 Phe-Arg-Arg-Leu-Ser(P)-Ile-Ser-Thr-Glu-Ser-Glx-Pro
Glycogen synthase (site 1b)[c]	10 Ser-Ser-Gly-Gly-Ser-Lys-Arg-Ser-Asn-Ser(P)-Val-Asp-Thr-Ser-Ser-Leu-Ser
Troponin I, rabbit heart	? Val-Arg-Arg-Ser(P)-Asp-Arg-Ala-Tyr-Ala
Histone H1, calf thymus	3 Ala-Lys-Arg-Lys-Ala-Ser(P)-Gly-Pro-Pro-Val-Ser
Type II regulatory subunit of cAMP-dependent protein kinase[c]	Asp-Arg-Arg-Val-Ser(P)-Val

[a]Taken from [2] (see also Chapters 1, 2, 7 and 10)

[b]The numbers above the phosphorylated residues refer to the rates at which the sites are phosphorylated relative to the β-subunit of phosphorylase kinase (100%) at a 6 μM substrate concentration under a defined set of assay conditions. Each protein was from rabbit skeletal muscle unless otherwise stated.

[c]Hvang, T.S., Feramisco, J.R., Glass, D.B. and Krebs, E.G. (1979) Miami Winter Symp. 16, 449–461.

258

enzyme on or off, but as a mechanism for interconverting an enzyme between two (or more than two in the case of multiple phosphorylations) different states which respond differently to substrates and effectors.

3. Synchronization of different metabolic pathways

An important generality which has been suggested by the systems discussed in this book, is that enzymes involved in biodegradative pathways are activated by phosphorylation, whereas enzymes involved in biosynthetic pathways are inactivated by phosphorylation (Table 2). This finding has clear implications for the co-ordinated control of cellular functions since it would allow different metabolic pathways to be regulated by the same protein kinases and protein phosphatases. This is already established in the case of cyclic AMP-dependent protein kinase which phosphorylates a large number of enzymes which are related functionally. The activation and inactivation of glycogen phosphorylase and glycogen synthase respectively allows co-ordinated control of the pathways of glycogen breakdown and synthesis in response to adrenaline (Chapter 1). The activation and inactivation of triglyceride lipase and glycerol phosphate acyl transferase (Chapter 4) allows co-ordinated control of triglyceride breakdown and synthesis by adrenaline. The inactivation of both glycogen synthase (Chapter 1) and acetyl-CoA carboxylase (Chapter 3) allows co-ordinated control of glycogen and fatty acid synthesis in response to glucagon.

The information presented in this book has also suggested that the actions of protein phosphatase-1 and inhibitor-1 are also not confined to glycogen metabolism. Protein phosphatase-1 has been implicated in the control of acetyl-CoA carboxylase (Chapter 3), glycerol phosphate acyl transferase (Chapter 4), cholesterol esterase (Chapter 5), and HMG-CoA reductase and reductase kinase (Chapter 6). The possibility that this enzyme is a major target of insulin action is discussed in Section 5.

A further protein which has multiple actions in a variety of systems is the calcium binding protein calmodulin (see Section 4 below).

It should however be emphasized that if two or more enzymes are phosphorylated and dephosphorylated by the same protein kinase and protein phosphatase, the different activities can still be regulated independently through substrates and allosteric effectors (Section 2). The existence of protein kinases and protein phosphatases which have multiple actions is therefore a device which allows for both synchronous and independent control of the various enzymes on which they act.

TABLE 2

Enzymes found in the cytoplasm of mammalian cells that are regulated by phosphorylation

	Chapter	Types of protein kinase involved			
		cAMP	Ca^{2+}-calmodulin	Other	
Activation by phosphorylation					*Biodegradative pathway*
Glycogen phosphorylase	1	−	+	−	Glycogenolysis
Phosphorylase kinase	1	+	+[a]	−	Glycogenolysis
Myosin	7	−	+	−	ATP hydrolysis
Triglyceride lipase	4	+	−	−	Triglyceride breakdown
Cholesterol esterase	5	+	−	?	Cholesterol ester hydrolysis
Inactivation by phosphorylation					*Biosynthetic pathway*
Glycogen synthase	1	+	+	+	Glycogen synthesis
Acetyl-CoA carboxylase	3	+	−	+	Fatty acid synthesis
Glycerol phosphate acyl transferase	4	+	−	−	Triglyceride synthesis
HMG-CoA reductase	6	−	−	+	Cholesterol synthesis
Initiation factor eIF-2	8	−	−	+	Protein synthesis

[a]Phosphorylase kinase phosphorylates itself at a slow rate

Pyruvate kinase (Chapter 2) has been omitted from this list since it can be regarded as a biosynthetic enzyme channelling glycolytic intermediates to ATP and fatty acid synthesis, or as a biodegradative enzyme involved in glucose and glycogen catabolism.

4. Calmodulin and cellular regulation

Although the importance of calcium ions in cellular regulation has been recognised for many years, the concept of an intracellular calcium receptor which mediates the actions of this divalent cation, in a manner analogous to the regulatory subunit of cyclic AMP-dependent protein kinase, is a very recent development. It is only since the discovery of calmodulin and the identification of its multiple actions that this idea has started to gain widespread acceptance [4–6].

Calmodulin, formerly termed the 'modulator protein' or 'calcium dependent regulator protein' was originally discovered as a small thermostable calcium binding protein which stimulated the activity of one of the cyclic nucleotide phosphodiesterases in brain [7,8]. This protein was subsequently found to be present in very large amounts in brain and other tissues (50–500 mg per 1000 g tissue), well in excess of the concentration of cyclic nucleotide phosphodiesterase. The amino acid sequence of calmodulin, which demonstrated its considerable similarity to troponin-C [9–11] was really the stimulus which started the search for other proteins whose activity might be affected by calmodulin. The seven enzymes which are known to be either activated or completely dependent on calmodulin are given in Table 3. In all these cases calmodulin is only an activator in the presence of calcium ions. This list includes three protein kinases, myosin kinase (Chapter 7), phosphorylase kinase (Chapter 1) and tryptophan hydroxylase kinase [16]. In addition membranes isolated from a variety of tissues have been shown to

TABLE 3

List of enzymes that are activated by or completely dependent on calmodulin

Enzyme	Tissue	Reference
High K_m cyclic nucleotide phosphodiesterase	Brain, heart	7, 8
Adenylate cyclase	Brain	4, 12
Calcium-magnesium ATPase	Erythrocyte membrane	13, 14
NAD-kinase	Higher plants	15
Phosphorylase kinase	Skeletal muscle	Chapter 1
Myosin kinase	Skeletal muscle and smooth muscle	Chapter 7
Tryptophan hydroxylase kinase	Brain	16

contain endogenous calmodulin-dependent protein kinases which phosphorylate a variety of membrane proteins whose identity and function are unknown [17]. Calmodulin has been shown to become associated with the mitotic spindles during cell division [18,19] and to stimulate the depolymerisation of microtubules [19]. It seems likely that many more calmodulin-dependent enzymes will be identified over the next few years, and that its involvement in other cellular processes, such as secretion, will become recognised.

It seems, however, already justified to describe this protein as a major intracellular calcium receptor, and the calcium-calmodulin and cyclic AMP–cyclic AMP-dependent protein kinase systems are closely analogous as illustrated in Figure 1.

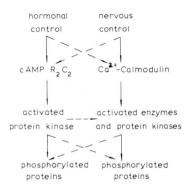

Figure 1. Role of cyclic AMP-dependent protein kinase and calmodulin in the neural and hormonal control of intermediary metabolism

When cells are stimulated by agents which elevate the intracellular concentration of cyclic AMP, cyclic AMP binds to the major cyclic AMP receptor in the cell, which is the regulatory subunit of cyclic AMP-dependent protein kinase. The active catalytic subunit then dissociates and phosphorylates those substrate proteins which are present in that particular cell. Similarly, stimuli which elevate the intracellular concentration of calcium ions, cause calcium ions to bind to the major calcium receptor in the cell, which is calmodulin (or troponin-C in the specialised case of the muscle contractile apparatus). The calcium-calmodulin complex then associates with a number of enzymes, some of which are protein kinase. This leads to the activation of the protein kinases and to the phosphorylation of their substrate proteins.

The major difference between the two regulatory pathways is that mammalian cells contain just two cyclic AMP-dependent protein kinases (type I and type II). These isoenzymes contain different regulatory subunits, but have the same catalytic subunit which phosphorylates a variety of intracellular substrates. On the other hand, the current evidence suggests that there are a large number of different calmodulin-dependent protein kinases each of which have very restricted substrate specificities. Nevertheless, both mechanisms produce the same end result, namely the phosphorylation of a number of proteins in response to either cyclic AMP or calcium ions.

Another important point illustrated by the broken lines in Figure 1 is that cyclic AMP and calcium-calmodulin pathways are closely interlinked and may often be interchangeable in different tissues. Thus the stimulation of glycogenolysis in skeletal muscle is a β-adrenergic effect mediated by cyclic AMP (Chapter 1) whereas in the liver it appears to be largely an α-adrenergic effect mediated, at least in part, by calcium ions [20]. Similarly, electrical stimulation of skeletal muscle leads to an elevation in the concentration of calcium ions in the muscle sarcoplasm, whereas electrical activity of the brain can elevate the intracellular level of cyclic AMP. The latter effect could however be a consequence of the activation of the calmodulin sensitive adenylate cyclase, which is present in brain but not apparently in most other tissues (Table 3).

A second point at which the cyclic AMP and calcium-calmodulin pathways may interconnect is at the level of the protein kinase. Figure 1 suggests that cyclic AMP-dependent protein kinase may often phosphorylate calmodulin sensitive enzymes. Two examples of this phenomenon were described in this book, namely the activation of phosphorylase kinase in skeletal muscle (Chapter 1) and the inactivation of myosin kinase in smooth muscle (Chapter 7).

A further interconnection may also occur at the level of the substrate proteins for cyclic AMP-dependent protein kinase. Figure 1 suggests that cyclic AMP-dependent and calmodulin-dependent protein kinases may often phosphorylate the same proteins, although at different sites, due to the high specificity of protein kinases. Three examples of this phenomenon have already been described; glycogen synthase (Chapter 1), a brain membrane protein, termed protein-1, which is believed to be involved in synaptic transmission [21], and phospholamban [22], the protein in cardiac sarcoplasmic reticulum believed to be important in the uptake of calcium ions from the sarcoplasm (Chapter 7).

5. *The mechanism of action of insulin*

Although it has been known for over 50 years that insulin lowers the concentration of glucose in the blood, attempts to elucidate its mechanism of action have so far met with little success. Insulin not only stimulates glucose uptake but a range of biosynthetic processes including glycogen, fatty acid, steroid and protein synthesis. Its mechanism of action surely represents the most important outstanding question in the hormonal control of intermediary metabolism. The following hypothesis is based on the original observations made by Joseph Larner and his associates in glycogen metabolism, on the information presented in Chapters 1, 3, and 6, and on some very recent work. It offers the hope that many of the actions of insulin will be explainable in terms of a single basic mechanism.

In the late 1950s and early 1960s Joseph Larner and his associates made several important observations that provided the framework for subsequent work on insulin action [23,24]. Firstly they found that glycogen synthase, which catalyses the rate limiting step in glycogen synthesis, was activated by the allosteric effector glucose-6P. Secondly, they noted that while the activity of the enzyme in muscle extracts was identical in the presence of excess glucose-6P, the degree to which the activity could be stimulated by glucose-6P depended on the metabolic state of the cell. In particular, when rat hemidiaphragms were incubated with insulin the activity of glycogen synthase measured in the absence of glucose-6P was increased and this correlated with the increased rates of glycogen synthesis. The additional finding that the effect of insulin was still observed when the hemidiaphragms were incubated in the absence of added glucose indicated that the effect was not a consequence of an increased rate of entry of glucose into the muscle in the presence of the hormone [25,26].

The subsequent finding that glycogen synthase could exist as a dephosphorylated form which was almost fully active in the absence of glucose-6P, or as a phosphorylated form which was largely dependent on glucose-6P clarified the situation, since it indicated that the interaction of insulin with its target cells had either decreased the activity of glycogen synthase kinase or increased the activity of glycogen synthase phosphatase [27].

In 1969–1970 cyclic AMP-dependent protein kinase was identified as a glycogen synthase kinase [28,29] and at this time it appeared that many of the actions of insulin were going to be explainable in terms of decreases in the intracellular concentration of cyclic AMP. This view was based on a number of experiments that had been carried out in adipose tissue and in liver. Insulin was shown to lower the concentration of cyclic AMP in adipose tissue that had been previously elevated by adrenaline and it also inhibited the rate of lipolysis that was stimulated by adrenaline [30] (Chapter 4). Similarly in the liver, insulin decreased the concentration of cyclic AMP that had been previously elevated by glucagon, and it also inhibited the output of glucose that was stimulated by this hormone [31]. However in skeletal muscle or diaphragm, insulin could not be shown to decrease the concentration of cyclic AMP in either the presence or absence of adrenaline [32,33]. Moreover in adipose tissue and in liver, subsequent work demonstrated a lack of correlation between the effects of insulin on cyclic AMP concentration and its effects on metabolism. Thus adenosine produces a much greater decrease in the concentration of cyclic AMP than insulin in adipose tissue, but has no effect on the rate of lipolysis, while in the presence of adenosine, the further addition of insulin strongly inhibits lipolysis without any further decrease in the concentration of cyclic AMP [34]. Similarly in rat liver the effect of adrenaline on hepatic glucose output is largely on α-adrenergic effect which is unrelated to changes in cyclic AMP concentration [20]. The antagonism of this effect by insulin is therefore also unlikely to be related to cyclic AMP. The effects of insulin on cyclic AMP therefore appear to be secondary to some more primary action of the hormone.

The realisation that glycogen synthase can be phosphorylated by at least three different protein kinase (cyclic AMP-dependent protein kinase, phosphorylase kinase and glycogen synthase kinase-3; see Chapter 1) further complicated the situation, since it raised the question of which glycogen synthase kinase might be under the control of insulin. The recent finding that insulin activates glycogen synthase in the soleus muscle of I-strain mice which lack phosphorylase kinase activity [35], appears to exclude the possibility that insulin activates glycogen synthase by decreasing the activity of phosphorylase kinase. It also implies that the action of insulin on glycogen synthesis is unrelated to decreases in the intracellular concentration of calcium ions, or to a direct effect on the calcium-calmodulin regulatory pathway (Figure 1).

In order to avoid the complication of multiple phosphorylations, we recently investigated the effect of insulin on the state of phosphorylation of inhibitor-1 (Chapter 1), since this protein is phosphorylated on a single threonine residue by just one protein kinase (cyclic AMP-dependent protein kinase). Using a perfused rat hind limb system, insulin was found to decrease the degree of phosphorylation of inhibitor-1 2-fold [36]. This result would appear to narrow down the search for the target enzyme for insulin action to either cyclic AMP-dependent protein kinase or to the protein phosphatase which dephosphorylates inhibitor-1 in vivo. The latter enzyme is likely to be protein phosphatase-1 (see Chapter 1).

The recent studies on the control of steroid synthesis by hormones (Chapter 6) appear to be of some significance to our understanding of insulin action. Ingebritsen and Gibson have shown that insulin promotes the dephosphorylation of both HMG-CoA reductase and HMG-CoA reductase kinase in isolated hepatocytes. Since neither HMG-CoA reductase kinase nor HMG-CoA reductase kinase kinase are cyclic AMP-dependent protein kinase, the action of insulin in this system does not appear to be explainable in terms of an inactivation of cyclic AMP-dependent protein kinase.

The striking finding that emerges from the effects of insulin on glycogen synthase, inhibitor-1, HMG-CoA reductase and HMG-CoA reductase kinase, is that insulin promotes a net dephosphorylation of each of these enzymes. Furthermore, protein phosphatase-1 has been implicated as an important enzyme in the dephosphorylation of all of these proteins. In addition, it should be mentioned that acetyl-CoA carboxylase, the rate limiting enzyme in fatty acid synthesis, is also activated by insulin (Chapter 3). Since acetyl-CoA carboxylase is inactivated by phosphorylation, it seems likely that the action of insulin on this enzyme will ultimately be explainable in terms of a net dephosphorylation of this enzyme. As discussed in Chapter 3, protein phosphatase-1 has been implicated as a major acetyl-CoA carboxylase.

The foregoing discussion has suggested that many of the actions of insulin on intermediary metabolism could be explained if protein phosphatase-1 became activated in response to insulin. The mechanism by which this might take place has been suggested by the recent results of Goris, Merlevede and coworkers. These workers reported some years ago that liver contained a phosphorylase phosphatase which was only active in the presence of ATP and magnesium ions [37]. This 'ATP-Mg dependent phosphorylase phosphatase' was recently resolved into two components termed F_A and F_C [38]. F_C was an inactive phosphorylase phosphatase, while F_A was a factor which activated F_C in the presence of ATP-Mg. This system has now been demonstrated to exist in skeletal muscle and cardiac muscle as well as liver [39], and accounts for a substantial proportion of the total phosphorylase phosphatase activity of these tissues [39].

Although the ATP-Mg dependent phosphatase was originally believed to be specific for the dephosphorylation of phosphorylase, recent work in this laboratory carried out in collaboration with Dr. Jozef Goris has shown this is not the case. The enzyme also dephosphorylates phosphorylase kinase (β-subunit), glycogen synthase, acetyl-CoA carboxylase, HMG-CoA reductase and HMG-CoA reductase kinase. Furthermore the rate at which each substrate is dephosphorylated relative to phosphorylase is identical to that observed with protein phosphatase-1. In addition, the ATP-Mg dependent phosphatase is inhibited by the same concentrations of inhibitor-1 that inactivate protein phosphatase-1 [40]. These observations raise the exciting possibility that the ATP-Mg dependent protein phosphatase (F_C) is merely an inactive form of protein phosphatase-1. The relationship between F_C and protein phosphatase-1 may therefore be perfectly analogous to that which exists between cyclic AMP-dependent protein kinase and its catalytic subunit. In this case, the inactive F_C component would be composed of a catalytic subunit, C, and a regulatory subunit, R, while protein phosphatase-1 would merely contain the catalytic subunit, C. Since the activation of F_C by F_A requires ATP-Mg, F_A may well be a protein kinase. The activation of F_C might then involve phosphorylation of R by F_A. In this hypothesis, the activity of F_A is controlled by insulin, while the susceptibility of the activated protein phosphatase to inhibitor-1 allows the effects of insulin to be cancelled by hormones that raise the intracellular concentration of cyclic AMP (Figure 2). Since the concentration of insulin in the blood is $10^{-9} - 10^{-10}$M, whereas the concentrations of F_A and F_C are likely to be in the range of $10^{-6} - 10^{-7}$M, the effect of insulin cannot be direct. The interaction of insulin with its specific receptors on the outer surface of the plasmic membrane would therefore have to trigger the formation of an intracellular activator of F_A (Figure 2). Another attractive feature of this hypothesis is that F_C might not be the only intracellular protein that is activated by F_A, and that F_A like cyclic AMP-dependent protein kinase, may have multiple functions. The next few years should prove or disprove these ideas.

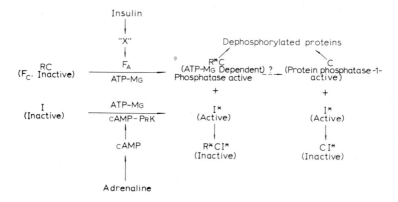

Figure 2. Hypothesis for the mechanism of action of insulin. Abbreviations: cAMP-PrK, cyclic AMP-dependent protein kinase; I, inhibitor-1, 'X', postulated second messenger for insulin; R,C, regulatory and catalytic subunits of protein phosphatase-1; R*,I*, phosphorylated derivatives of R and I. F_A and F_C are defined in the text.

6. *Protein phosphorylation as a regulatory device*

Although this book has mainly been concerned with the regulation of enzyme activities and with the control of the major pathways of intermediary metabolism, perhaps the majority of hormonal effects are those which influence processes where the regulatory proteins are not enzymes. These include effects on muscle contractility (Chapter 7), on secretion, on membrane transport and permeability, and on protein induction and degradation. It seems likely that the hormonal regulation of many of these processes involves protein phosphorylation, but progress is hampered by the lack of understanding about the molecular nature of the processes themselves. Chapters 8, 9 and 10 dealing with the initiation of protein synthesis, ribosomal function and chromosome division, clearly indicated the technical difficulties associated with the study of complex systems in which the phosphoproteins involved are not enzymes. The lack of suitable in vitro assays have so far prevented the role of protein phosphorylation from being understood completely. The finding that the phosphorylation of histone H1 takes place at a precise time during the cell cycle and the suggestion that it is a signal for chromosome division (Chapter 10) is particularly interesting, since it raises the possibility that the basis of many circadian and diurnal rhythms may be the reversible covalent modification of certain proteins.

Protein phosphorylation is also involved in the binding of divalent cations [41]. The clustering of phosphoserine residues in the egg yolk protein phosvitin and the milk protein casein are likely to have important functions in binding iron and calcium respectively. Phosphoproteins also play important roles in bone formation [41]. Finally, the recent discovery that the transforming factors of tumour viruses are protein kinases [42] is an exciting development which implicates protein phosphorylation in viral infection and cancer. A detailed discussion of this topic can be found in Volume 2 of this series [43].

References

1. Nimmo, H.G. and Cohen, P. (1977) Adv. Cyc. Nuc. Res., 8, 145–266.
2. Cohen, P. (1978) Curr. Top. Cell. Reg. 14, 117–196.
3. Cohen, P. (1979) Control of Enzyme Activity, Chapman-Hall Ltd., 11 New Fetter Street, London, EC4P 4EE.
4. Wolff, D.J. and Brostrom, C.O. (1979) Adv. Cyc. Nuc. Res., 11, 28–288.
5. Cheung, W.Y. (1980) Science, 207, 19–27.
6. Klee, C.B., Crouch, T.H. and Richman, P. (1980) Ann. Rev. Biochem., in press.

7. Cheung, W.Y. (1970) Biochem. Biophys. Res. Commun., 38, 533–538.
8. Kakuichi, S., Yamazaki, R. and Nakajima, H. (1970) Proc. Jap. Acad., 46, 589–592.
9. Vanaman, T.C., Sharief, F., Awramik, J.L., Mendel, P.A. and Watterson, D.M. (1976) in: Contractile Systems of Non-Muscle Tissues (Perry, S.V. et al., eds.), pp. 165–176, Elsevier/North-Holland.
10. Dedman, J.R, Potter, J.D., Jackson, R.L., Johnson, J.D. and Means, A.R. (1977) J. Biol. Chem., 252, 8415–8422.
11. Grand, R.J.A. and Perry, S.V. (1978) FEBS Lett., 92, 137–142.
12. Brostrom, C.O., Huang, Y.C., Breckenridge, B.M. and Wolff, D.J. (1975) Proc. Natl. Acad. Sci. USA, 72, 64–68.
13. Gopinath, R.M. and Vicenzi, F.F. (1977) Biochem. Biophys. Res. Commun., 77, 1203–1209.
14. Jarrett, H.W. and Penniston, J.J. (1977) Biochem. Biophys. Res. Commun., 77, 1210–1216.
15. Anderson, J.M. and Cornier, M.J. (1978) Biochem. Biophys. Res. Commun., 84, 595–602.
16. Yamauchi, T. and Fujisawa, H. (1979) Biochem. Biophys. Res. Commun., 90, 28–35.
17. Schulman, H. and Greengard, P. (1978) Proc. Natl. Acad. Sci. USA, 75, 5432–5436.
18. Welsh, A.J., Dedman, J.R., Brinkley, B.R. and Means, A.R. (1978) Proc. Natl. Acad. Sci. USA, 75, 1867–1871.
19. Dedman, J.R., Brinkley, B.R. and Means, A.R. (1979) Adv. Cyc. Nuc. Res., 11, 131–174.
20. Exton, J.H. and Harper, S.C. (1975) Adv. Cyc. Nuc. Res., 5, 519–532.
21. Huttner, W.B. and Greengard, P. (1979) Proc. Natl. Acad. Sci. USA, 76, 5402–5406.
22. Lepeuch, C.J., Haiech, J. and Demaille, J.G. (1979) Biochemistry, 18, 5150–5157.
23. Villar-Palasi, C. and Larner, J. (1960) Arch. Biochem. Biophys., 94, 436–442.
24. Craig, J.W. and Larner, J. (1964) Nature (London), 202, 971–973.
25. Villar-Palasi, C. and Larner, J. (1961) Biochim. Biophys. Acta, 139, 171–173.
26. Danforth, W.H. (1965) J. Biol. Chem., 240, 588–593.
27. Friedman, D.L. and Larner, J. (1963) Biochemistry, 2, 669–675.
28. Schlender, K.K., Wei, S.H. and Villar-Palasi, C. (1969) Biochim. Biophys. Acta, 191, 272–278.
29. Soderling, T.R., Hickinbottom, J.P., Reiman, C.M., Hunkeler, F.L., Walsh, D.A. and Krebs, E.G. (1970) J. Biol. Chem., 245, 6317–6328.
30. Butcher, R.W., Baird, C.E. and Sutherland, E.W. (1968) J. Biol. Chem., 243, 1705–1712.
31. Robison, G.A., Butcher, R.W. and Sutherland, E.W. (1971) in: Cyclic AMP, Academic Press, New York.
32. Goldberg, N.D., Villar-Palasi, C., Sasko, H. and Larner, J. (1967) Biochim. Biophys. Acta, 148, 665–672.
33. Craig, J.W., Rall, T.W. and Larner, J. (1969) Biochim. Biophys. Acta, 177, 213–219.
34. Fain, J.F. (1974) Biochem. Ser. One., 8, 1–23.
35. LeMarchand-Brustel, Y., Cohen, P.T.W. and Cohen, P. (1979) FEBS Lett., 105, 235–238.
36. Foulkes, J.G., Jefferson, L.S. and Cohen, P. (1980) FEBS Lett., 112, 21–24.
37. Goris, J. and Merlevede, W. (1974) FEBS Lett., 48, 184–187.
38. Goris, J., Defreyn, G. and Merlevede, W. (1979) FEBS Lett., 99, 279–282.
39. Yang, S.D., Vandenheede, J.R., Goris, J. and Merlevede, W. (1980) FEBS Lett., 111, 201–204.
40. Goris, J., Stewart, A.A. and Cohen, P. (1980) manuscript in preparation.
41. Weller, M. (1979) Protein Phosphorylation, Pion Ltd., London.
42. Collett, M.C. and Erikson, R.C. (1978) Proc. Natl. Acad. Sci. USA, 75, 2021–2024.
43. Hunter, A. (1980) in: Molecular Aspects of Cellular Regulation (Cohen, P. and VanHeyningen, S., eds.), Vol. 2, Elsevier/North-Holland, Amsterdam, in press.

Index

Cohen (ed.) Recently discovered systems of enzyme regulation by reversible phosphorylation
© *Elsevier/North-Holland Biomedical Press, 1980*